CATAPULT

CATAPULT

How to Think Like a Corporate Athlete to Strengthen Your Resilience

PUNIT DHILLON

TORPEDO
» PUBLISHING

CATAPULT
How to Think Like a Corporate Athlete to Strengthen Your Resilience

ISBN 978-1-5445-2187-9 *Hardcover*
978-1-5445-2186-2 *Paperback*
978-1-5445-2188-6 *Ebook*
978-1-5445-2192-3 *Audiobook*

For Nina,

My better half. You make me laugh, make the world's best chocolate chip cookies, challenge me in profound ways, and make me whole—you are the chocolate chips to my cookie!

CONTENTS

SECTION THREE: RISE UP

FOREWORD

by DR. ANNALISA JENKINS
Global Biopharmaceutical Leader and Champion for Diversity and Inclusion in the Field of Global Health

Over the course of my career, people have often asked me what I think constitutes a great leader. After considering this question time and time again, I developed a response I feel defines exemplary leadership while offering a nod to the inimitable Albert Einstein's famous equation, albeit slightly altered: EMC^2.

Leaders must be able to envision, engage, energize, enable, execute, measure, communicate, and collaborate. Great leaders have a clarity of vision about the future they want to create and they engage others to follow them on that journey. They create energy around their purpose while identifying and removing things capable of draining energy from the people around them. They enable their team to get work done every day, and every day, they get work done themselves. They regularly measure their own progress and they are clear communicators and enthusiastic collaborators.

Punit Dhillon represents the next generation of great leaders. Through his own leadership journey, Punit has learned that the ability to build great teams and make a real, sustainable impact is rooted in purpose—and he is not afraid of articulating that. Nor is he afraid of genuinely paying attention to people and finding ways to connect with and inspire them. As both an executive and an athlete, Punit does not shy away from pushing both his team and himself to reach peak performance because he knows there is no room for only "practicing" when you are trying to address the big issues of the world.

In these ways, Punit lives a life of thoughtful, measured fearlessness centred on strong values, unshakable authenticity, and passion for purpose. He works to be a source of light in his company by being constantly conscious of how his actions impact others and is committed to his community and the nonprofit sector.

Punit also understands leadership is not about a start, a middle, and an end—it is a process of continual, lifelong learning that requires hard work, deep reflection, strong values, and an innate desire to motivate those around you. This book is a practical guide to embarking on that journey, equipped with as much foresight and understanding as possible, and I can think of no better time for such a book to exist. At the time of this writing (Fall 2020), we are experiencing a leadership crisis. We are living in the middle of a global pandemic while simultaneously witnessing profound changes to the earth's climate. Challenges to the role of multilateral institutions have led to unpredictability and insecurity. The world needs to reset, and we know that shift is going to be founded upon the leadership of the next generation.

The best thing today's great leaders can do is coach and mentor those coming up behind them while building upon their strengths. The best thing the next generation can do is start the journey by taking the first few steps into the shadows armed with the experience of those who have already made taking on the world's challenges their mission. Punit is not afraid of those shadows. He is ready to share some of his own light with you and coach you along every step of the way.

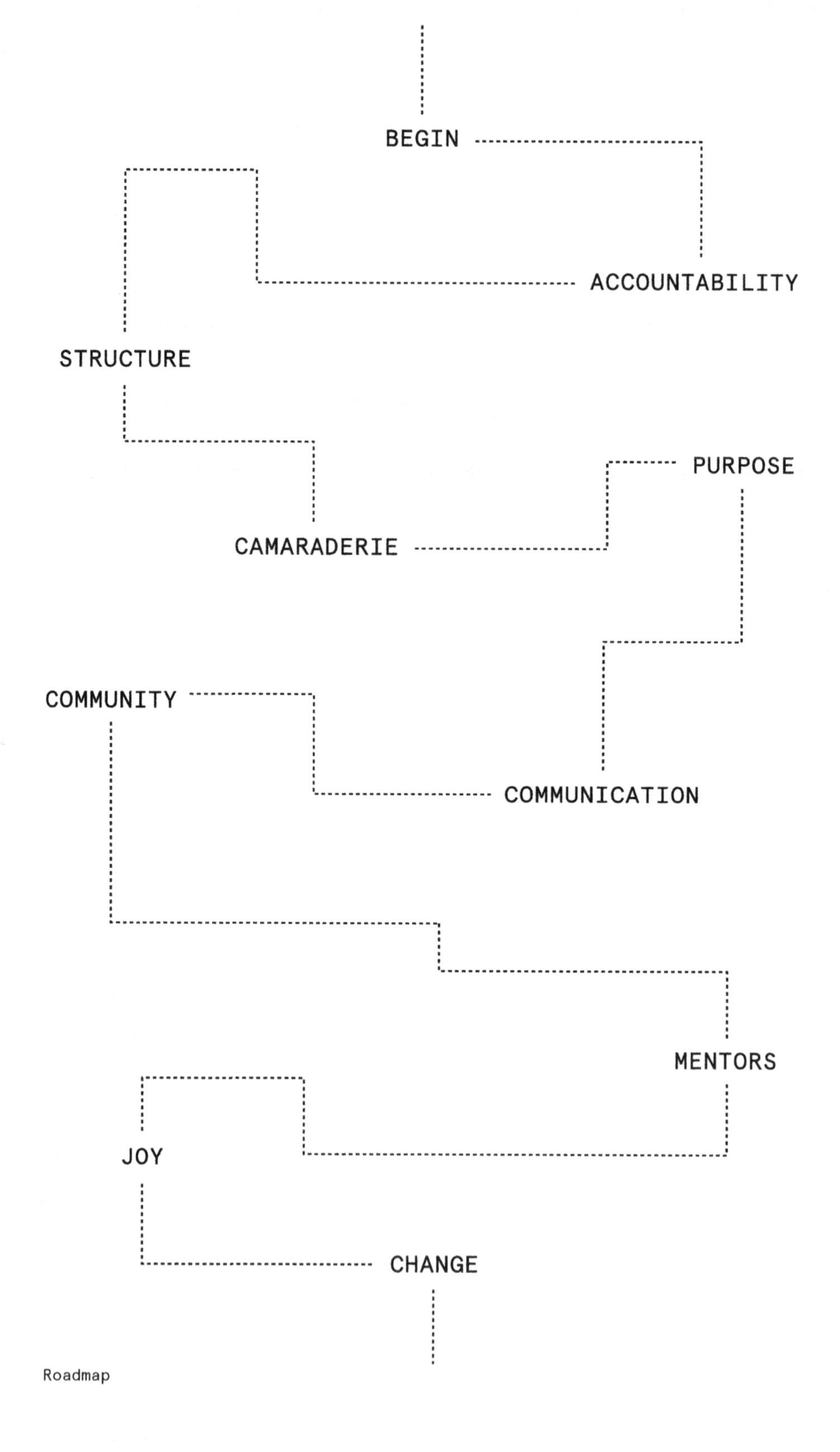
BEGIN
ACCOUNTABILITY
STRUCTURE
PURPOSE
CAMARADERIE
COMMUNITY
COMMUNICATION
MENTORS
JOY
CHANGE

Roadmap

PREFACE

"I am never going to make it to the finish line."

Hours into my first Ironman race, after swimming 2.4 miles, biking 112 miles, and running almost 26 miles, I had a mere 800 metres left in the race. I could see the end—it looked like a straight shot—but as with many things in life, the optics were deceiving. As the path softly bent toward the general direction of the finish line, blurry in the distance, it felt so close I could almost touch it. Then, ever so slowly, the road under my feet turned in the opposite direction, leading me away from the end of the race. I realized I still had a half mile to go. Everything I had worked for was right in front of me. The endless hours of training, the commitment, the sacrifice—it was all about to pay off. All I had to do was keep going for a few more metres, just one minute more, but I was convinced I could not do it.

Halfway through the marathon, pain had started radiating from my feet. The extreme heat of the day had prompted well-meaning spectators to fire up their garden hoses and mist us runners as we passed by, offering some much-needed refreshment while simultaneously creating perfect conditions for blisters. I was numb from the pain, but now that I was nearing the end, the agony became unbearable. Every step was excruciating.

Something about that final length of the race sparked a sense of defeat within me. Mentally, it stretched on for miles with no end in sight and I was stumbling along so slowly thinking I would never reach the end. But I kept going, step by step, second by second, until finally, unfathomably, I crossed the finish line.

By then, my body was in such shock, the end of the race did not feel real. I frantically asked the people around me where I needed to go for the next part of the competition.

"Nowhere," they said. "You made it!"

Several hours and some much-needed medical treatment later, I reflected on those last few metres. My wife had to replay the finish for me, so I could hear the famous words—*"You are an Ironman!"*—just to make sure I was not dreaming. It turned out that while I assumed I had been running at my slowest pace, I had actually been sprinting, the last mile being my fastest in the entire marathon. All my training had kicked in and allowed my body to carry me through to the end when I needed it the most. I had done the work and given myself a strong enough foundation that when things got rough, I was not only able to perform, but also do so at my best.

Once I was back in my right mind and realized what I had accomplished, absolute euphoria enveloped me. While it lasted, I felt invincible. I had extreme confidence in my ability to do anything and everything I devoted my time and attention to.

I have relived that race in my mind countless times. I come back to it any time I am experiencing hardship or questioning my capabilities. I think about it whenever I am doing anything to better myself or prepare for future challenges. When I returned to the office, the phrase *"Why not?"* became a persistent echo in my head. Everything felt like a possibility. I am told that when people undertake extreme endurance events such as the Ironman, they emerge with a similar mindset wherein every request, goal or dream feels within reach.

There are many parallels that can be drawn from the athlete to what I refer to as the Corporate Athlete (the super entrepreneur) in this book. There are many similarities in behaviour, training, and approach that both the athlete and the Corporate Athlete undertake in order to realize a dream. As an entrepreneur, I want to capture this feeling of undeniable athletic success and translate it to the workplace, boardroom, and life, so you do not have to complete an Ironman to experience the same joy, but you can use the same energy to catapult yourself to new levels of personal success, satisfaction, and contentment.

THE JOURNEY

In entrepreneurship, there is never a finish line. You are perpetually pushing yourself past any imaginary line you set for yourself, constantly training and working on becoming better, smarter, more efficient, and more effective.

In the business world, those seemingly endless last 800 metres will be the obstacles that are going to try to impede your path. They will be the big problems you are trying to solve. They will be the insecurities you will have to face. They will be the doubting voice circling inside your head. They will be every stakeholder, every project derailment, every missed opportunity. The only way to use these things as fuel to your fire is by knowing what you are capable of and living your true purpose.

Living this way takes courage, and in order to harness this confidence, you have to trust yourself. Real confidence arises not by a sense of bravado but by mastering the art of knowing who you are and what your purpose is in this life. These qualities will underline every action and decision you will make. With this clarity, you can approach your entrepreneurial adventures in a holistic manner, constantly recalibrating depending on the situation at hand. My intention is to arm you with the skill set that allows you to achieve your goals. When failure is not an option, tenacity and resilience will give you that confidence to help you succeed. The good news is I already know you have the drive. Just by taking the initiative to enter the demanding, daunting world of entrepreneurship, you are showing bravery, determination, and perseverance, all of which you will need as you continue along your career path.

At this point, you will likely feel an intense drive to tackle the significant, wide-reaching problems facing our world today. You will probably have great ideas and solutions to such problems as well as the energy and enthusiasm to execute them. But with a lack of experience, you inevitably will grapple with what most budding entrepreneurs must confront at the onset of their careers: a fear of failure.

Perhaps you fear the challenges you will face because no matter what field you are in or how experienced you are, they will come. Fear has the power to distract you, consuming your time and attention. If you let it, it can stop you before you even get started.

You are not alone in experiencing how daunting it can be to identify and then solve a problem. Being an entrepreneur *is* hard. But every significant contribution to our world involved challenges like those you will confront on your own journey. Everyone from Oprah Winfrey to Steve Jobs had to start somewhere, and none of their success simply fell into their laps. Oprah was unceremoniously removed from her co-anchor position for the WJZ 6 p.m. news program in Baltimore and was demoted to a lower profile position at the station. She has referred to this as her worst failure in her television career. Steve Jobs was famously ousted from Apple at the beginning of his career, only to return years later to resurrect the company and elevate it to its current innovative status. While Winfrey and Jobs were putting in the work that was required of them, a great deal of failure occurred along the way. Entrepreneurship is risky business, which means failure is a probability. The only way to cultivate bravery is by stepping up and taking the chance, much in the same way you pin the race number on your chest to prove to yourself everything you are capable of achieving. Success hinges on risk—you cannot have one without the other. It demands a mental toughness and confidence that only comes from looking inward first. Young entrepreneurs must physically and emotionally invest in themselves to prepare for the unavoidable challenges ahead. Doing so requires reflection, introspection, time, and effort. There are no shortcuts. As a young entrepreneur, you can tackle real problems, but only after you have put in the work to discover the kind of person, entrepreneur, and leader you want to be.

In these pages, I plan to give you the tools you need to do just that. As an unrelenting entrepreneur myself, I have distilled the lessons I have learned over the years to help you discover and understand your unique purpose, which will give you the confidence and perspective you will need to overcome any fears and challenges. I will show you how to tap into your purpose, to live and lead with authenticity, and to truly prosper, regardless of your chosen field, with endurance and stamina.

Some entrepreneurs have so much confidence, they feel they can skip steps and still make it to the mythical finish line. This is not the book for them. People who feel they already know all the answers are welcome to deal with the consequences of such a mindset. But if you are interested in digging deeper to develop yourself into the strongest, most resilient person you can be so you can successfully make a lasting impact on the world, then this book is for you.

The framework of this book is based on ten principles I believe are the basic tenets for entrepreneurial success. As an accomplished athlete, I have seen firsthand the benefits of implementing a regimented training plan in order to reach my goals. As an entrepreneur, the same concepts can apply to maximize your corporate performance—if similar formulas and training can be translated to the business world, there is nothing Corporate Athletes cannot achieve. In *Catapult*, I will map out these principles and refer to tangent experiences in mini sections called *Breakaways*, which, in cycling, refers to the moment when a competitor successfully opens a gap ahead of the peloton. The parallels and patterns between athletics and business will be easy to decipher and we will tackle the endurance journey together. The utilization of *Breakaways* throughout the principles will allow me to show you through various examples how I have had an unlikely journey in life, sports and business and how it has allowed me to find my own purpose. Whether you happen to be a marathoner, weekend jogger, swimmer, cyclist, young, old, in shape or not, you can do this. I know because I did it and I have witnessed others do it. This book will help you prepare for your metaphorical race and provide you with a plan for leading a purpose-driven corporate and professional life. Everything in these pages will help you tap into your true potential, as a person and as an entrepreneur.

Much of this work requires you to have accountability to yourself and to others while remaining true to yourself and your purpose. This is a concept most people do not fully address until they are deep into their careers. But without an understanding of how to remain authentic, both personally and professionally, entrepreneurs often lack the inner strength necessary to maintain alignment between their own vision and the demands of their stakeholders, resulting in constant pressure and unmet or unrealistic expectations. To better understand who you are as a human, as a leader, and as a person capable of affecting real change, you will have to identify your true values and ethics. You will have to define the underlying principles you believe in. You will have to learn to trust your own abilities and have a sense of authenticity so you can gain clarity around your true purpose.

As entrepreneurs, we have an innate desire to solve every problem we possibly can and to monetize every opportunity. Unfortunately, there just is not enough time to do so. As much as we wish we could, we simply cannot take on every issue facing our world today and at the same time, not everything has to be profitable to be fulfilling. This is why purpose is so critical. Understanding your own unique purpose provides you with the clarity necessary to pinpoint the specific impact you can make. Knowing who you are and what you stand for will inform you as to which challenges you should tackle. The more you are in touch with your own sense of purpose, the easier it will be for you to clearly communicate and accomplish your vision, whether that be with yourself, your team, your external stakeholders, your employees or the world.

The big question then becomes: How do I define my purpose so I can become accountable both to myself and to others? Purpose is essentially who you are and why you exist at this particular time in history. It is the essence of all the things you are meant to fulfill and all the changes you are meant to make in this world. It is a big ask, yet it is also deeply motivating. Purpose heeds you to answer the call—are you ready?

The principles detailed in these pages have served me well on my journey to self-accountability and it is my hope that in exploring them in this book, you will be on the fast track to understanding your own. Although these principles may not provide the answers to every obstacle you are going to encounter, they form a good basis that will inspire you to ask bigger questions. I encourage you to find what works specifically for you as you determine the best way to apply your own unique skills to solve global issues. There is no single recipe for success, no one, sure-fire solution to every hurdle you will need to clear. But these concepts can act as a roadmap as you work to identify your own guiding principles.

I will be the first one to tell you that I do not know everything. I have not reached a point in my career where I believe I have made it, or I am done learning or growing. I am hoping to offer an introspective view of my life and career so you can learn from my mistakes, my observations, my weaknesses, and my triumphs. I am particularly motivated to share all of this now because I believe there is a great deal happening in the world that demands the attention of big thinkers. Helping others succeed so they can tackle the world is a priority for me because I know, collectively, we can affect lasting change and positive growth.

If you are inspired to make strides in the world, I encourage you to focus on the following arenas as a starting point: education, life sciences, technology, artificial intelligence (AI), and the environment. These subjects will be covered at the end of this journey as fields that can challenge you as you look for solutions for bettering this world and advancing humanity toward a more hopeful future. Most importantly, it is a reference and opportunity to frame a series of public conversations that extend beyond this book and provide you with a call to action. Your willingness to pursue your purpose has the power to help shape decisions for technology leaders, policymakers, and citizens from all income groups, nationalities, and backgrounds.

UNRELENTING ENTREPRENEUR

I have always felt destined to tackle the entrepreneurial world from day one. I was born in Canada, then sent to India to live with my grandparents for the first five years of my life while my parents finished their educations. When I returned to Canada, I bore witness to my parents' tireless work ethic and soon learned they expected a great deal from me as well. They created a schedule for me that included time each day for academics, athletics, and reading. They even required I dress a certain way, which meant I would leave the house wearing dress pants, and unbeknownst to them, would change into shorts en route to school. They expected me to be home from school at a certain time, calculating how long it would take me to get from the school to the house and if I was even a few minutes late, I would expect an interrogation. In my youth, my dad would regularly take me to a park nearby to do hill repeats, climbing as fast as we could, again and again. I loved it. I would run those hills, propelled by the high of running and the encouragement from my dad. His coaching made the exercise fun and the experience evoked inside of me a drive that would eventually fuel my competitive spirit. I was nine years old when I finally beat him up the hill for the first time and it was my proudest personal achievement.

The disciplined fashion in which I was raised has manifested in different ways throughout my work as an entrepreneur, a career that has carried me through various ventures primarily in the healthcare field, culminating in launching my own biotech company focused on cancer immunotherapy at the age of twenty-nine. But it took many years lived and lessons learned before I could reflect on the principles that shaped my experience and led me to where I am today.

When I first entered the workforce, I was all about pleasing whomever I worked for as I was not yet clear on my own personal purpose. I had ambition and a plan to eventually go into corporate finance law, which I knew would require work and ladder-climbing, but that was the closest I came to having any kind of internal accountability.

My own work ethic served me well in my early days as an entrepreneur when my focus was entirely on doing whatever it took to meet the goals of my organization. I put everything I had into this mission, which was most of my energy and more hours than I could count. But I knew it all was necessary for the growth of the organization and, ultimately, my own personal growth as well.

I made countless connections in those early days, a requirement for gaining any headway when you are first starting out, and I soon realized I could only network effectively when I truly understood what I believed in and why I was doing the work in the first place. I was starting to realize just how important knowing and living your own purpose really is.

I lucked out and met some supportive senior entrepreneurs who were interested in my success and I worked hard to demonstrate my ability to handle responsibility and deliver results. I wanted to establish myself as a person of my word by always following up with what I said I would do.

My career trajectory carried me to a wide range of roles, including:

- Vice president of finance and operations at Inovio Pharmaceuticals, a biotechnology company focused on the synthetic DNA products for treating cancers and other diseases,
- Cofounder, CEO, and president of OncoSec, a biotechnology company pioneering new technologies to stimulate the body's immune system to target and attack cancer, and
- Board member and shareholder of Emerald Health Sciences, committed to enhancing health and well-being through diverse investments in pharmaceuticals, botanicals, and bioceuticals.

Throughout my career, I was able to build a reputation for being consistent, reliable, and dependable and that made people comfortable trusting me. I also realized how critical it was for entrepreneurs to find sources of support and guidance as they navigated their career paths, understanding my own responsibility in helping those who were following in my footsteps. This would eventually lead to the launch of YELL Canada, a registered charity I cofounded with two of my closest friends, that provides school districts with cutting-edge curriculum delivered through teachers, guest speakers, and workshops. The primary focus of YELL is on the foundational learning of business as well as leadership, self-development, and mindfulness within the business world.

I have always found the more genuine you are and the more you surround yourself with people that help reinforce that quality, the better off you are. Because, ultimately, any success you will experience—at work, at home, in all areas of life—comes down to being sincere. As you will soon learn in the pages of this book, living with authenticity starts with accountability to others and, above all, to yourself.

BREAKAWAY

BEGIN

"A journey of a thousand miles begins with a single step."
—Lao-Tzu

You begin every journey you take in your life by simply starting. Every step you take is a seed you plant that will lead to being held accountable to something. Sometimes, it can really be something as literal as a planted seed that blooms into a berry.

My grandparents owned a blueberry farm in the outskirts of Vancouver and growing up, my brother and I were often tasked to pick blueberries every summer. It was an onerous undertaking. My grandmother would affix a sizable basket around my waist, four to six of which could fill up a larger twenty-pound bucket of blueberries. On a good day, I was expected to pick an entire row of blueberry bushes, which was no easy feat. This amounted to over 200 pounds of berries and we were paid by each twenty-pound pail we picked. As a child, I would stand at the top of the row of bushes, squint in the sun, and wonder how and when I would ever get to the end. Somehow, I always did—you simply begin.

With the sun beating down on my back, I would pull a bushel of berries with my hands and drop it into the bucket. I had to be fast, yet thorough. I had to have quick reflexes, scanning the bush for ripe fruit, and moving nimbly. After all, I was paid by the bucket and if I wanted to see the fruits of my labour, there was no time to waste.

Once the buckets were full, I would run over to the larger collection bins to empty them and hurry back. My mom and grandmother would enthusiastically encourage us from the sidelines. My father, in the meantime, would challenge us kids during these picking sessions, seeing how many blueberries my brother and I could pick in an hour, pitting us against each other. This friendly competitive spirit was one that I picked up during those summers and I carry over those traditions to present day, regularly challenging my daughters to relay races. Any seemingly insurmountable task can be injected with some fun, otherwise, the days can feel very long. Berry by berry, the buckets would be filled and I would move from one bush to the next. Like clockwork, by the end of the day, I would have completed the entire row of blueberry bushes, my fingers stained a deep purple. What seemed impossible only a few hours ago, was now accomplished. It was good and done.

I am a recreational triathlete, but I did not become one overnight. I started by running up a single hill, not knowing what was on the other side. Today, I compete in any race that feels like a challenge.

Much like blueberry picking, or anything in this life that you want to accomplish, you can think and dream about it all you want, but there comes a time when you physically need to take that first step. There is no other way. Over time, that single blueberry and single step all add up to something much bigger. You can do anything if you want to. You just have to show up and begin.

A WORLD-CLASS APPROACH

Growing up, athletics was the heartbeat of my daily life, and its profound influence permeates each principle in this book for good reason. In my youth, I was a competitive swimmer. Today, I am a runner, cyclist, and two-time Ironman triathlete. During my training, I have often reflected on the demands placed on world-class athletes. They spend an inordinate amount of time training and improving so they can be prepared for what appears to be the relatively few times they are called on to perform. Soon, they reach the off-season and, after an average of about twenty years or so, depending on the sport, they retire.

Compare that to the life of a world-class Corporate Athlete, a term that can apply to any top-level executive dedicated to performing at the height of their potential. They work forty, sixty, even eighty hours a week. They barely get three weeks of vacation and, even then, they are constantly interrupted. There is no such thing as an off-season. They do it for four or five decades or more.

How can you get a Corporate Athlete to perform like a world-class athlete? Professional athletes are expected to deliver consistently. They are especially expected to be at their best in do-or-die situations. The same expectations are placed on Corporate Athletes, but they often lack the one thing keeping professional athletes performing at their peak: resilience.

Resilience is one of the key qualities desired in leaders today, but many people confuse it with toughness. Toughness is certainly an aspect of resilience, as it may enable an individual to separate emotion from the negative consequences of difficult choices. It can be advantageous in the business world and often celebrated, but only to a point. It can create an armour that deflects emotion and can cut you off from the resources you need to bounce back. Resilience is not about deflecting challenges, but about absorbing them and rebounding stronger than before.

I will be touching on many personal experiences that culminate in the principles I use in this book, which are meant to offer you a framework to apply or adapt in order to define your own set of principles. The structure of these principles is crafted in a way that allows you to accept setbacks as they occur, move on, and create new possibilities.

If you are an entrepreneur who feels like you are in over your head, being pulled in too many directions, and losing sight of why you started the work in the first place, you likely lack resilience. You may not fully understand your true purpose in this world. Maybe you are trying to take on too much yourself and not seeking help from others. You might only feel accountable to people outside yourself. The good news is you can change all of that, and it is my hope the principles explained in this book will show you how.

This book will teach you to become the best leader and Corporate Athlete possible by helping you discover your true purpose, so you can gain the confidence to tackle the problems impacting our world today. It will teach you how the positivity you create by helping others can come back tenfold. It will show you how taking on these issues requires a collective effort that starts with your own ability to look outward and how working with others in your community who are invested in your success reinforces what you believe in. It will show you how creating accountability within yourself results in others trusting you and your vision. The structure will fortify your resilience, strengthen your work ethic, infuse your life with more joy, and, in some instances, improve your physical performance. You will learn how all of this works together to perpetuate your purpose, propel your drive, and catapult you toward the goals you have for your work, your life, and the world around you.

GAME FACE

SECTION ONE

It starts and ends with us. Before we form our communities, before we look to others, we only have ourselves to answer to. It is in this space of solitude where we make the vow of self-responsibility. Only then, can we be confident enough to brave this great big world and make our mark on it.

PRINCIPLE 1

TRUE ACCOUNTABILITY

"Change will not come if we wait for some other person or some other time. We are the ones we've been waiting for. We are the change that we seek."
—BARACK OBAMA

When I learned I had to complete a mass swim at the start of my first triathlon, I was fairly confident. Because I was a great swimmer in a pool as well as a trained lifeguard who had completed similar ocean drills, I felt

I would naturally be a good open-water swimmer. At the beginning of my training, however, I misjudged just how difficult this task would be and was rightly put to the test each session. But with enough time devoted to the practice, I soon transitioned from false confidence to true confidence, as my training leading up to the race had built up my resilience for multiple open-water swims in the lake, the bay, as well as the ocean.

On one of my very first ocean-training swims, the coach asked us to swim past the break, swim back, and run up a hill for six repetitions. It seemed reasonable, easy even, based on my athletic history and my overconfidence. However, I was wearing a new wetsuit and it was not yet broken in, the neckline constricting and chafing me each time I took a stroke. After the second round, I already felt out of my depth; the waves were so intense, much more so than I was expecting. I made the mistake of trying to swim over them instead of under them, and if the timing was off, I was pummeled by the water crashing down on me with brute force. In addition to battling the elements, I was restricted not only by the unforgiving wetsuit, but also by extreme shortness of breath. What did I sign up for? Why did I think I could do this? Why did I think I was going to excel? Pushing myself to try to stay ahead of the pack, beating everyone in and out of the water and up and down the hill for no apparent reason was self-defeating, especially when it was not a race. I was not as prepared as I thought I was for training and because nature had its own ideas, both factors triggered me into a fight-or-flight reaction where my body shut down midway through the workout. I could not hear anything but a deafening hum. I could not take a deep enough breath to fill my lungs. I could not see beyond the glare reflecting off the water, refracting into a million tiny dots. My brain thought I was ready; my body did not. It was a frightening and uncontrollable feeling, but one I needed to experience.

Months later, I stood on the shores of Okanagan Lake in British Columbia with no waves or grievances to worry about, aside from a small hole I had to repair in my wetsuit. What would it be like to swim with almost 2,000 people? My imagination was running wild. I knew I had worked hard and if I kept calm and stuck to my plan, I would come out of the water without any trouble, yet I was feeling everything but prepared. I had been in mass swim starts before, but usually for shorter races, and the anxiety I felt at that exact moment, seeing all the swimmers on the shoreline waiting for the race to begin, was nerve-racking. Thankfully, standing next to me was my brother, Maheep. He was also racing with me and we had done a lot of the training together, including a tune-up race a few months prior. He helped to calm me down. He reminded me of the race plan, that there was no point getting anxious or overly excited for the mass swim start, and to just create some space for myself and swim my swim. This was easy for him to say. He is 6'1" and 220 pounds—no one was going to swim over him or pull on his legs. I knew this was going to be a long swim and I needed to conserve energy. Maheep reminded me that I had nothing to prove, and everyone was trying to do the same thing. Inside, I knew this, but it was nonetheless reassuring to hear. I had the training under my belt; I just had to trust in it and stay focused on myself, not anyone else.

Unconsciously, we all tend to pick up the pace to keep up with others. But what if different people are swimming for different reasons? The word "sports" almost always seems to imply a competitive element, which means we are constantly comparing ourselves to someone else. Competitiveness is an important force in life. It is what drives the market and is behind some of the world's most impressive accomplishments. On an individual level, however, it is critical that you know whom you are competing with and why, and that you have a clear sense of the space you are in.

World-class athletes know they are responsible for their own level of success. Only you know the race you are running. Whether you are a world-class athlete or world-class Corporate Athlete, it is on you to do the hard work and put in the hours of training to reap the rewards. If you are a professional athlete, this may be a world championship; if you are a Corporate Athlete, this may be the valuation of your business. Each one of us has a unique potential and purpose and we are the only ones who can evaluate and set the terms of our lives and our success. Often, we are impacted by other people and make their approval the standard we feel compelled to meet. As a result, we squander our very potential and purpose. Remembering what your goals are and keeping them in your sights will be your anchor in times when comparison or competition threatens to pull you away from your path.

In a team, you have a responsibility to bring your best to every performance so you can win together. If you participate in a solo sport, you know the weight of every victory or failure lands squarely on your own shoulders. Marathon runners, for instance, might have a coach or other runners they train with, but at the end of the day, they know it all comes down to them. If they are not holding themselves accountable, they are never going to conquer a personal best, let alone cross the finish line. Entrepreneurship is a marathon. To succeed, you have to have the endurance, discipline, and confidence to keep going, mile after mile, race after race, and the only way to have the stamina and strength necessary to keep going and keep improving is to find accountability within you first. You will always have external stakeholders to answer to. They will expect certain things from you, and you will have to deliver. But the responsibility you have to those people is secondary to the accountability you have to yourself.

Lack of personal accountability leads to a lot of wasted time, energy, and money, which is exactly what every entrepreneur wants to avoid. We have to be efficient. We cannot waste money, especially when dealing with shareholders who have entrusted us with their hard-earned capital and are looking for a return on their investment. Everything must be about performance, growth, and results. When you are accountable to yourself first, it leaves little ambiguity for everyone else because you are in alignment with what you can deliver.

According to Seneca, the Roman Stoic philosopher, the Greek word *euthymia* is one we should think of often: it is defined as the sense of our own path and how to stay on it without getting distracted by all others that intersect it.[1] In other words, it is not about beating the other guy. It is not about having more than others. It is about being who you are and being as good as possible at it, without succumbing to all the things that draw you away from your own personal formula for success. It is about going where you set out to go and accomplishing the most that you are capable of in what you choose. If you need to put a value around that measurement, I have an equation to help you practice that personal accountability for success. It is time to sit down and think about what is truly important to you and then dismiss the rest. Without this, success will not give you the sense of pleasure you seek, leaving much more to be desired.

1 Giovanni A Fava, Per Bech, "The Concept of Euthymia," *Psychotherapy and Psychosomatics*, November 27, 2015; 85(1): 1-5. doi: 10.1159/00441244. PMID: 26610048.

THE EQUATION

What is accountability? I believe you can break it down into this simple equation:

$$\text{ACCOUNTABILITY} \times \frac{(\text{VALUES} + \text{TRUST} + \text{STRUCTURE} + \text{MOTIVATION}) \times \text{POSITIVE HABITS}}{\text{INTUITION}}$$

True Accountability Formula

Before you can take on accountability to a new boss, team, company, board or stakeholder, it is critical that you take the time to really think about what each of these elements means to you. There is no shortcut around it. Each one is important and your thoughts and actions around these concepts will determine how you do what you do, why you do it, and how you align yourself with everyone around you.

VALUES

Values are the litmus test determining your priorities. They influence your behaviour, define your character, and show you how closely you are to living the kind of life you want. What matters most to you? What morals do you consistently abide by? What are your intentions? What do you want your character to stand for?

When you think critically about these questions, you begin to understand how your values affect everything you do, big and small. You will be clear in the decisions you make because you will have a better understanding of what is informing your choices.

Whenever I am in my hometown of Vancouver, I run the same five-mile loop that starts at the home we have lived in since 1985. I pass the elementary school and the spot where my dad used to take me on hill repeats, looping around Queen Elizabeth Park, where Percy Norman Swimming Pool once stood. Here, I am reminded of witnessing Mark Tewksbury's record-setting 100-metre backstroke swim in a short-course pool. I continue past McBride Annex, where I almost failed kindergarten because I spoke English poorly, and pass the home where I was born. I run by the Dairy Queen, where I ate many Dip Cones as a child, and I finish along Victoria Drive, where I have witnessed the transformation of a traditional main street filled with stores owned and run by first-generation immigrants, now scattered among bustling spots for hipsters and foodies. I run this route to anchor myself and remind myself where I came from. This is the city that shaped me; these are the people and experiences that made me who I am today. Each experience has cemented my values and informed who I wanted to be as a leader.

I am lucky to have the ability to return to my hometown on a regular basis so I can make regular check-ins with myself. Without fail, every single time I come home, it is an exercise in peeling back the layers to reveal who I am essentially. When you take away the responsibilities, the career, the current life you live, who are you at your core? The city may constantly change, but I am still the same person I was growing up—this is always a good reminder. I understand that not everyone can go home all the time. But even without physically being there, you owe it to yourself to consistently check in and ask yourself the hard questions: Who am I? What do I stand for? What matters most to me? What do I believe in? This will ultimately ensure that the values you have are in agreement with the decisions you make.

TRUST

The trust you must have in yourself stems from your belief in your own level of competency. While entrepreneurship can require some degree of risk, no one should jump headfirst into any venture they are not sure they can handle. You will be hard-pressed to find an athlete attempting a 100 km ride on their first day of training. There needs to be an intrinsic sense of knowing they can do it one day, but this needs to be balanced with their level of ability. Likewise, entrepreneurs need to be aware of their own skills, knowledge, and experience, all of which inform them to pursue their objectives. Trust in its most basic form reaffirms your ability, integrity, and confidence, which ultimately encourages you to explore new possibilities and ways to mindfully push the envelope. Trust inherently centers on your character and your ability to say what you mean and do what you said you are going to do, which can at times mean a willingness to admit when you do not have all the answers. Such cases might require you to seek out mentors or experienced people in the field who will guide you to the places you want to go. At the end of the day, this sense of trust naturally extends beyond the individual level and parlays into your behaviour in relation to others.

At OncoSec, we decided to make trust a priority by taking a page from Stephen Covey's *The Speed of Trust*, which dissects thirteen behaviours required for creating and maintaining trust, including practicing straight talk, demonstrating respect, offering transparency, and righting wrongs. We reviewed these behaviours, zeroed in on those that resonated most with us, and created actions around them in order to have more trusting relationships throughout the organization. For instance, if I was not spending enough time with my general counsel, I would pick one of these thirteen behaviours and then spend a few minutes demonstrating it, which could include anything from showing loyalty, working on listening first or clarifying an expectation.

After doing this for some time, I realized fostering trust mostly comes down to communication. How do you communicate with an individual to confirm you are both on the same page? How can you ensure, through communication, that mutual trust is being established and respected? Because that is how a strong relationship is formed between two people. The bigger the organization, the more work is required for establishing those relationships. You want to reinforce those habits down the chain throughout the entire organization, leading by example, because when you do, even a new employee can feel the same level of trust and understanding because it is simply a part of the company's culture. Trust becomes especially crucial in a time when many companies are pivoting to work-from-home scenarios. This became abundantly apparent during the COVID-19 pandemic, which occurred during the writing of this book. Companies were forced to trust that their employees were capable not only of performing, but also thriving in extreme circumstances, and as a whole, the world's workforce responded and delivered. New norms were instituted as people proved they were up to the task of remaining productive and continuing to contribute, even in the most unpredictable of times. Reinforcing mutual trust in these instances is fundamental to ensuring that the same rules of communication remain unchanged, regardless of the employee environment. Understanding that this is the way of the future is key for companies to evolve along with changing times.

Building trust takes time. We did these exercises regularly throughout the day and practiced in workshops. We could not just spend a few minutes here and there and expect lasting change. We wanted to create sustained trust, and that meant our entire management had to be committed to focusing on a continued effort toward achieving it.

Another crucial element of trust is the ability to deliver on promises. When you fall short on delivery, you only confirm that your word and your actions do not line up. When others see you delivering on promises, they will give you more trust incrementally as you continue to prove your skills. Just like marathon training, earning trust requires consistency. The more trust you earn from others, the more you will trust in yourself. This feedback loop is integral to your overall personal and professional success. The key is to recognize that it begins with you—before you can ever trust in others, you have to first trust in yourself.

STRUCTURE

Structure is the factor that keeps you aligned with your purpose and provides you with a framework for performing at your best. In a company, structure can include corporate governance, meeting schedules, and routines, all of which work to reinforce whatever needs to happen so everyone can reach their individual and shared goals. In athletics, it is the training plan you create with your coach that you need to follow in order to get the outcome you are looking for. Structure keeps you focused and moving forward so you never lose sight of your goal.

Structure requires repetition of your routine so that, over time, everything becomes automated, and you can devote more of your time to focusing on the things you want to pursue. You know what time you are going to get up every day. You know when you are going to work out. You know when you will have time with your family. You never have to waste energy thinking about the minutiae of your day because it is all pretty much mapped out for you. This automation is meant to maximize productivity and creativity. Everybody has a structure to their day, however organized or chaotic it may appear. This routine is your fallback, the one thing you can count on when you feel overwhelmed or lost.

We often witness grace, poise, and an overall sense of ease of action in individuals but may not take a moment to recognize the underlying reasons as to how they make it look so effortless. When you observe an athlete in motion in any sport, it often looks spontaneous and perfect—you never really see the culmination of hours of frustration, practice and refinement. An elite athlete can make the performance of such skills look easy due to their dedicated discipline. Similarly, a Corporate Athlete's success depends on and is the direct result of repetitive practice within a structured routine. The ease in the way you lead is due to the countless hours you have methodically invested in yourself, your skills, and your objectives.

Conversely, occasionally challenging your structure by disrupting your routine and introducing new components will also keep things from becoming too rote. Some people call this a cheat day, when they can do, eat or experience things outside of the norm. I like to refer to this as "conscious disruption," an interruption in the daily structure that could be the catalyst for many new and inspirational ideas. After all, a solid routine is wonderful 99 percent of the time, but you still need to make some room for discovery and invention.

MOTIVATION

When you start to slowly fine-tune your purpose, you will start to feel an intrinsic motivation to pursue it. The passion you feel around the work you are doing will inspire you to hit the ground running. There can be external driving factors as well, but ultimately, enjoying and loving the things you do can provide you with the motivation necessary to complete little tasks that add up to your big goals.

After all, you are capable of creating a currency for joy in your life—that currency is the time you devote to things that infuse you with positivity. I call this your Joy Quotient, and we will dig deeper into this concept in Principle Nine.

Your motivation is directly linked to your purpose. Sometimes, that might mean crazy days with intense travel schedules, long hours, and endless to-do lists. You might find yourself constantly hustling, always asking, "What comes next? How are we going to get there? What are we going to do to get there?" But when you understand your purpose, you know all of that hard work will pay off in the end.

That being said, motivation does not always come easy. If you are in a state of survival or experiencing stress, finding motivation can be difficult. During these times, I try to remember the joy, the runner's high (and the entrepreneurial equivalent) that I always experience to get me going. When things get really tough, I wholeheartedly subscribe to the bargain-reward system, where I bribe myself to complete tasks with the promise of a chocolate chip cookie at the end of the day. The use of sense memory and rewards are very effective when you find it hard to pull yourself out of a phase of apathy. I also firmly believe that motivation is linked to the idea of momentum and inertia, which means whatever you do, big or small, will bring you one step closer to where you want to go. Much like running, motivation and forward motion kick in the minute you take that single step. Sometimes, that is all you need to get the wheels turning.

In time, the more connected you are with your values, trust, and routines, the easier it will be to tap into your own motivation. When you are firmly connected to these elements, you will experience increased fulfillment and will become more motivated to accomplish your goals.

POSITIVE HABITS

An old adage says it takes twenty-one days to form a new habit. I like to compare the idea of positive habits to the macroeconomic principle of the multiplier effect, which refers to the proportional increase or decrease of final income, which is dependent on the amount invested or withdrawn. In other words: you simply get out what you put in. Working on this concept will require consistent dedication, but the payoff is invaluable. Positive habits reinforce every building block of the overall equation, whereas negative habits lead to self-sabotage. When you adopt positive habits that align with your values, trust, structure, and motivation, you are amplifying the effect each building block has on your true accountability. Much like drinking enough water with electrolytes during training to ensure your hydration levels are on par, giving yourself enough time at the end of a training session to cool down and stretch to prevent injury or meal planning to optimize the ratio of carbohydrates, protein, and fat you are consuming to ensure peak performance, adopting positive habits require consistent repetition. There are unlimited ways you can incorporate positive habits into your daily life and when life gets busy, you need to find innovative ways to work them into your routine.

There have been times where I have laced up my shoes and hit the pavement without a solid running plan. I like the idea and the feeling of being free and, some days, I dislike running in a perpetual loop around the neighbourhood. In these instances, I run where my heart takes me, and when I near the end of the allotted time, I map out the closest supermarket and I head over to pick up some things for the household. Afterward, I call my wife and she will drive over to pick me up and we head home. Is this an unusual approach to sticking to my daily runs? Absolutely. But this is just one example of how your approach does not have to look the same every day. It just needs to get done. On days when it feels like I cannot fit something in, I think outside of the box and come up with a creative way to make it happen. There are no rules here—you make them. When there is no room for excuses, how will you show up for yourself?

INTUITION

Intuition is an innate skill but one we need constant practice with to tap into and sharpen. I like to think of it like a muscle we need to constantly flex and place under pressure so it can perform when we require it. When something does not feel right, our brains and bodies have unique ways of letting us know. There needs to be an underlying thread of intuition behind every concept in this equation for true accountability to shine through. Our values, trust, structure, and motivation can all be refined with this intrinsic element of intuition so we can sense if we are straying too far off course. The idea is to have this intuitive compass so we can use it and rely on it to bring us back to true north.

In business, every entrepreneur will have a story about how they ignored their intuition and paid a big price for it, myself included. What I can tell you is there is no way to avoid the learning curve when it comes to listening to your own gut. The only way to work this muscle is to consistently check in with yourself. Am I talking myself into or out of a decision? Have I given myself enough time to really sit with the choices to feel out which makes the most sense to me in this moment? Am I leaning toward one decision because it is more sensible, or does it pay to be a little riskier in this instance?

There will be mistakes made and lessons learned, I guarantee it. But the sooner you learn to check in with yourself at every turn, the more attuned you will be to your intuition and the better you will be able to navigate your business decisions. You will never be lost again.

BREAKAWAY

THE CENTRAL GOVERNOR THEORY

The central governor theory of energy regulation in the body is based on the idea that your brain will override your physical ability to perform and will shut things down before permanent damage occurs. This is the primitive mode of communication between the body and the brain. It is why the final mile in a marathon is usually the hardest, but yet, the fastest—the body system does not know how much longer you have to go and your brain will naturally try to restrain you. Similarly, if you have a community around you, you tend to exercise harder when there are people around you than when you are on your own. When you are by yourself, it is very easy to be won over by your self-talk, whereas being accountable to another person gives you the sense of soldiering on and of not giving up so easily.

Being accountable to a system gives you the confidence to trust that system because you know the road ahead. You know your brain is going to try and talk you out of those long miles somewhere in the middle of the marathon. Knowing this and persevering in spite of it because you understand why this is happening, is key to pushing through plateaus. In those moments when your brain is telling you to stop, you can recognize that it is normal and it is all a part of the bigger picture.

PERSONAL VS. SOCIAL ACCOUNTABILITY

Accepting accountability requires continued and dedicated effort. It demands you to stay focused on each element of the equation in order to maintain a clear vision of your goals.

Once a system of accountability is established, your discipline will result in more lasting, rewarding experiences. You will understand how all the pieces of the equation work together and their proven ability to yield results will empower you.

The accountability formula is one I have worked with over the years and it has proven to be the golden ticket to keeping me consistent with the things I want to achieve. I encourage you to take what works for you and add and eliminate components so you can come up with your own formula that will bring you the most success. There is no perfect formula to get you where you need to go. Use the one outlined in previous pages as inspiration to get you started and improvise to make it your own.

Once you have established your baseline, start to think about what it means to extend accountability to your team. The idea of social accountability is important because it emphasizes the element of human connection. There is a sense of behaviour reinforcement when you become accountable to people other than yourself. We all know that when all hands are on deck, it becomes easier to achieve your goals. The journey from self to others is imperative because it takes the focus outward and we ultimately need to look outside of ourselves and around our community to see what needs to be done.

So why do you do what you do? That is the question every Corporate Athlete needs to answer. Only then will you understand what matters and what does not. You can say no and decide to opt out of a race or personal competition that simply does not matter in the bigger picture. You are free to ignore competition with other people because most of the time, they are not really competition at all. It is here where you can develop that true accountability to establish the resilience and the quiet confidence that Seneca refers to. Find out why you are after what you are after. Ignore those who interfere with your pace, who pull on your feet while you swim. When you turn inward and tune out the noise, your purpose and the accountability you have with yourself will suddenly become loud and clear.

□

WORLD-CLASS CODA

- Everything comes down to one—one atom, one cell, one person. Before we look outward, we need to look inward. How we orient ourselves in this life is a direct reflection of how we behave. Once we establish true accountability to ourselves, we can use this as a starting point to take the next step out into the world.

- Creating an accountability formula is integral to gaining clarity, with each component holding you up to a particular standard. This formula is intentional—it gives you a reason to wake up every single day, to do everything you want to do. This is your anthem—how do you want to show up?

- True accountability is one of the key components to fully knowing yourself and what you stand for. Once you have a firm grasp of this, confidence inevitably follows. With confidence, you can start to develop yourself into a genuine leader.

CATAPULT FORWARD

- Thinking back, are there instances when you have let yourself down? Is there a predictable pattern that created a similar outcome? Conversely, can you think of a time when you achieved a long-awaited goal? What steps did you take to ensure that you would get there?

- If you could draft your own accountability formula, what would it look like? Why have you included certain components and what do they mean to you?

- How do you keep yourself accountable in everyday life? What are the rituals that you do? How do you foster self-encouragement as opposed to self-discouragement?

There is an inherent power in community. When we are challenged by our peers to go beyond our capabilities, to do better and be better, and when we are bolstered by the same people, there is no ceiling to our collective success. In this life, while it is true that we can accomplish much on our own, we can accomplish so much more together.

PRINCIPLE 2

NEED A COACH?

"Our chief want in life is somebody
who will make us do what we can."
—RALPH WALDO EMERSON

"Go! Go! Go! Come on!" Coach Andrew Currie shouted from the deck of the community centre pool. The sun had not yet risen but the pool, once quiet only moments before, was now the centre of much fervent activity. I could hear his commanding voice cut through the commotion, the sound of my own breath, and the sound of our limbs, entering and exiting the water. That morning, I had stared at the whiteboard proclaiming my sentence for the day: a 10 x 200-metre butterfly set. For roughly the next forty-plus minutes, I would be subjecting myself to an intense act of physicality that would push my body to the limit and test my speed, endurance, and drive. This was no one's idea of fun, but at least the entire team would be undergoing this extreme exercise together.

Moments before we began, I watched the red second hand tick on the swimming clock as each swimmer leapt off the swimming block one at a time, five seconds apart. About twenty seconds in, the lack of adequate oxygen coupled with the extreme physical exertion caused me to feel nauseated, disoriented, and winded. For the next forty minutes, it felt like I was swimming with a piano strapped to my back.

Why was Coach Andrew making us do this? I wondered, although I really already knew the answer. This was not just a test of our skills. This was a test of how well we could work together. After all, if one person failed in this task, we all failed. If we succeeded, we succeeded together.

Coach pushed us on every stroke along the way, his voice acting like a beacon from the side of the pool, calling us home to the finish line. Even though it seemed like an incredible undertaking, none of us was capable of, Coach Andrew's ability to speak directly to each of us made it clear to our team exactly what we needed to get done. We were in this together, and his confidence in everyone's ability relayed confidence in every individual's own ability. A coach like Andrew gave no false impressions and he demanded full buy-in from the team because he practiced what he preached. He was completely dedicated to the sport, helping each athlete become the best they could be. So we pushed forward, metre by metre, breath by breath, on and on. We each rose to the challenge, not only because we had to, but also because we wanted to.

Elite swimming is notoriously a large time commitment and a lot of hard work. In the sports world, everyone practicing at the elite level probably theorizes that their sport is the hardest, and they all do have elements that make them extremely difficult. But swimming, truly, is one of the most difficult sports because of the physical demands around water resistance, breathing, technique, and fear. Why fear? Because in swimming, quitting is not an option—quitting means drowning. You cannot just quit swimming, the way you can quit a marathon or walk off from the court or the field. You are forced to finish, whether you want to or not, and that is something you understand before you ever set foot into the pool. At the elite level of swimming, getting over the fear of putting it all on the line and pushing your body to perform at a new level to get to your personal best takes a lot of courage. Swimmers are at their best when they are working hard and seeing the results of their dedication. The power of coaching is about extending that resilience to recognize that not every meet is a championship meet and improvement at that level does not happen all at once. In fact, the better you get, the harder it will be to improve.

In athletics, training programs are designed to incrementally challenge the athlete over time as a test to see how much improvement can be made. When you are young and new to a sport or skill, you improve by leaps and bounds, not only because you are learning at an accelerated rate, but because you are hyperfocused. You are also physically growing and changing quickly. The older you get, the slower you grow, and the better you are, the harder it will be to hit those personal bests. Muscles are meant to be placed under strain and constantly challenged in order to be pushed to their utmost limit. Depending on how much you are doing, you will improve at a certain rate, but it will be at your own rate. The rate at which you improve is going to be different than others, so you need to be wary of comparing yourself to your peers.

When it comes to the Corporate Athlete, the same rules apply—an entrepreneur needs to understand that in order to be successful time and again, a similar kind of provocation needs to occur in order to stay on the path of personal and professional betterment. Muscle pliability refers to the phenomenon of adaptation. With time, practice and training, an athlete can conform to different stressors to recalibrate in different conditions and remain agile in various environments. My intention is to apply this very principle to your behaviours and skills in the corporate world, so you can remain flexible in order to adjust to ever-changing business conditions.

Just as an athlete retains a coach to help them polish their skills, entrepreneurs need a guide to do the same, to confront and redirect them so they can be pushed to a new level of discomfort in order to make their accomplishments that much more rewarding. Think of the coach as an integral part of the success of your training—much like how selecting the perfect shoes for your feet will help improve your performance, finding the right person to guide you is imperative to your corporate success. Once you have started to identify your purpose as a Corporate Athlete, the most clear-cut path to knowing who you are, what you stand for, and what you have been put on this earth to achieve is realized by working with a coach. Coaching can give entrepreneurs and leaders the structure and tools necessary to perform to the best of their ability. Coaching challenges you to learn, holds you accountable, reaffirms your commitment, and creates a framework in which you are continually improving. It provides the support you need to shoot for the stars while having a structure in place to help you navigate obstacles.

Without the guidance, structure, and support my teammates and I had received from Coach Andrew, none of us would have been able to push through that 10 x 200-metre set, which we all did together as a team. He was resolute in his support of us and his confidence in our abilities, and that translated to success in the water. With the right support, any task can be accomplished. After you take that first step to begin a journey, you have to be open to being coached.

GROWTH

Before you can ever benefit from anything a coach is able to offer you, you have to put your ego aside. You must be willing to let your guard down, open up, and grow as an individual before you can effectively lead a team. If you are constantly worried about what everyone else is doing, trying to follow the trends, and attempting to copy your competition, you will inevitably be catering to your ego. Working with a coach allows you to plan and prepare. Instead of being reactive, you become proactive in terms of what you want to achieve. It might take a few tries to find the person who is right for the job, but there is somebody out there who is capable of pushing you, elevating you to the next level, and working with you to think actively. This is essential in moving you forward into unchartered territory.

We all have room to grow and can find people who can help us do it. Some people are very lucky and work with one coach for most of their lives—Bob Bowman and Michael Phelps are one such example. Other times, you may need to find new mentors once you have outgrown them, recognizing the fact that you need to constantly reevaluate where you are in life and how much further you want to reach. But no matter how convoluted the journey can become, it starts with the realization that accepting help is a sign of strength, not weakness. Guidance allows us to learn through other people's experiences, both failures and successes, enabling us to improve leaps and bounds and at a faster rate via their expertise. Hard work is important, but it cannot be the only thing we rely on to become successful.

BREAKAWAY

CATALYSTS FOR GROWTH

When you observe and analyze the careers of successful athletes and nonathletes, one thing will be clear: success is not merely a function of what each person did, but a result of the significant impact others had on his or her success. Before winning his first of six NBA world championship titles with Coach Phil Jackson, Michael Jordan was coached by Dean Smith at the University of North Carolina, and Doug Collins, the head coach of the Chicago Bulls. They were both instrumental in recognizing his talents and developed game plans and systems to maximize Jordan's strengths, which allowed him to flourish as a prolific scorer and be recognized as the best player in the NBA, well before any championship title.

In the corporate world, the most important "other" person is often the boss. Certainly, I can attest to this in my career. We live in an era where we believe we can do anything and everything by ourselves, but the truth is a lot of professional success people experience is not just because of what they did on their own, but because of the quality of bosses they had at the beginning of their careers. Just as an age-group coach makes a meaningful impact on the career of an athlete, I believe that long-term career success is especially linked to the quality of your boss in your foundational years, but coaches will continue to impact you profoundly throughout the duration of your career. Consider Serena Williams—when she started, her father was her coach, and he was able to propel her to incredible heights. Later, after experiencing a health scare and losing her first Grand Slam, Williams began working with Coach Patrick Mouratoglou and has experienced many victories since. The right coach at the right time can make all the difference.

Generally speaking, there are broadly two types of bosses. The first type is one who is focused on getting results out of his or her subordinates. Their primary orientation is to get the job done, and they follow up, support, and drive you to get the results required for the organization. They leverage your talents and experience to build structure that is organizationally focused, with little attention paid to personal development. The second type is equally committed to delivering results, but also learns and builds on your experience while doing so. This boss prioritizes results but also focuses on asking you questions in a way that makes you reflect, thereby pushing you to get to the solutions yourself, rather than giving you the answers. They build the reflection habit by asking uncomfortable and challenging questions such as, "What could you have done better?" They are the catalysts in converting the time you spend at work into a positive experience, continuing to layer and reinforce your individual growth *within* corporate growth. More importantly, they establish the framework for measurable individual growth and foster results toward your purpose for the rest of your career.

The former type of boss ensures that you deliver the results—you might even get a promotion—but they do not aid your conversion of time into meaningful experience because the work is task-oriented. They are not enablers in establishing the structure to ensure alignment with your purpose for deeper personal and professional growth. They have not made a permanent positive difference to you. It is my hope that you will cross paths with the latter type of boss, one who not only will encourage you within the corporate framework, but also nurture your growth, setting you up for future success. When a boss is able to catapult you to new heights, they can also catalyze real personal growth within you.

Success is dependent on many factors in addition to hard work, and coaching and mentorship is key to navigating your career as it evolves over time. There is always more work to be done and it is never about a start and a finish—it is about finding ways to push past your finish line.

In my field of life sciences, when you think you have eradicated a disease with your drug, the objectives change and there is still more to do. For example, beyond curing a disease, there are opportunities for lowering the drug cost or improving the response rate or effectiveness. You will not always know the next best step or how to handle every scenario and you will need someone to help you continue to grow as the scope of your work evolves. If you want to be a world-class Corporate Athlete and leader, it is imperative for you to establish a plan and work with a coach who helps you stretch past your objectives, setting the standard for your team.

FINDING MY COACH

There is a psychological phenomenon called the Pygmalion effect, which is based on the idea that high expectations lead to improved performance in a given area. It is closely linked to the concept of a self-fulfilling prophecy. When pursuing any goal, you might think you are an expert from the get-go and you very well may become an expert eventually, but in between, there exists a learning gap you must address, either through experience or relying on the expertise of others. Naturally, a great way to start is by finding a coach.

The importance of coaching was not always clear to me. When I first started my company, I did not necessarily subscribe to the idea of corporate coaching because I did not see the potential or the need. It is also hard to distinguish who is genuinely an accomplished and suitable coach. Thankfully, I was referred to someone who had entrepreneurial, business, and coaching experience. She had a reputation for being direct and exhibited radical candour. Although I knew of her for years, I formally met Judy Brooks in January 2014, and she helped me create a set of tools and the framework that became the key to defining my purpose. Before I had those tools, I felt like I was operating in the blind. I was very motivated to work hard and I had a plan for my life but understanding why I am wired the way I am gave me a deeper understanding of my purpose and how I could use it as a lens and filter for everything I do.

We first connected so she could help me improve my speaking skills. We hit it off over the phone and a week later, she invited me to a weekend event she was organizing for a dozen other executives at Brew Creek in British Columbia. "I cannot tell you much about it," she said, "but I think you are going to get a lot of benefit out of this."

I was intrigued.

"Okay," I said. "Sign me up."

A few days later, eleven other male executives and I showed up at Brew Creek, which was in the middle of the woods. There was no cell phone coverage, but it would not have mattered if there was. Judy had a strict rule: no devices. There was a small cabin with a fireplace and, at first, we all just hung out, making pleasantries. But soon it was time to get to work.

Judy told us we each had to complete a five-minute presentation before dinner answering one question: Who is a person who has had a positive influence in your life?

The group was made up of people of different ages, demographics, and industries, but everyone opened up, sharing stories of the people who have shaped them into who they are today. I spoke about my uncle and his unique ability to both challenge and elevate me, much like a coach does an athlete. Something about the remoteness of the environment and the small, intimate group encouraged us to have these immersive, revealing conversations, which only got deeper and more meaningful over the next three days.

Leading into the weekend, the most intense part consisted of completing four different measurable personal profile assessments: behaviour patterns, emotional intelligence, ethics, and motivational factors. Each of these assessments is used to recognize and apply thinking, personality, and behaviour patterns to help identify and link said attributes in order to understand and personally define one's purpose. The data supporting each of the specific results from the assessments represents a clear plan to allow individuals to have recognizable relationships and gain an understanding of themselves between different facets of life, including work, home, community, family, and education. As you can imagine, there are hundreds of different types of assessments measuring different areas of your life, from personality to preferences. Outside of doing standardized testing for school and a few cognitive tests related to my career, this was the first time I had taken part in such a comprehensive internal review, and even this can be considered just brushing the surface. For those who are assessment neophytes or who have not had the opportunity to partake in personal profile assessments, I have provided information on the ones that I have taken in case you would like to explore further on your own. Note that each of these assessments includes a report and one-to-one debrief, and the results need to be interpreted by a certified administrator (you can find a brief description of a few of these assessments in the appendix).

Each test took about an hour and the questions are meant to be repetitive to ensure accuracy from these responses. There was no way to cheat the system. Once we were done, we dove into our individual results. We shared our different life experiences, both personal and professional, and the things that influence us most. By now, we had established a fraternity among us. We had been spending every minute together doing the same things every day, and anyone who was a stranger on day one was a friend by day three.

We all spent time digesting our individual evaluations and then we headed home. Three months later, we came back together for another three days, during which we took all the learning we did about ourselves during the first weekend and used it to craft a purpose statement. I spent a lot of time developing mine, but in the end, it came down to a very simple line: *My purpose is leading people and communities to pioneer indelible impact.*

I have always been interested in doing innovative things, and at that time, I was working in a cancer immunotherapy company that I cofounded. I certainly felt like we were making an impact with technology that could help save the lives of people who were suffering from stage four melanoma. But if I had not been doing that work, I still knew I was motivated by giving energy to others to elevate them in some way. I am that person high-fiving people on my morning run. I am the guy knocking on my neighbour's door after dinner, saying, "Hey, do you want to get some dessert?" I am constantly looking for opportunities to build or strengthen my community. It is more than a job because it is what motivates me every day and informs everything I do.

Now that I knew my purpose, Judy's job was to reinforce it. I hired her as my executive coach, and we started to talk almost every day. She would remind me of the things I needed to spend time on and point out opportunities for areas where I could grow. We would discuss my company's milestones and how I was working to meet them. If I knew I had to have a tough conversation with someone that day, she would offer suggestions for how to handle it. Together, we built a structure of support around my purpose, and it helped me become a better leader.

The experience was so impactful, I ended up taking my entire executive team to Brew Creek to undergo similar assessments with the same facilitators. I invested in having access to those facilitators for my entire management team for one year. I wanted each member of my team to do the same work I did because I knew the more explicit they were about their purpose, the more they would be aligned with their roles and responsibilities in the company. We set aside time quarterly to gather as a team and pick something to work on: behaviours, goal setting, anything that would continue to amplify us as an organization but that was also established in individual purpose and accountability.

The payoff, both personally and from a business standpoint, was immense. During that phase, the company had its biggest growth. We went from a $20 million market cap company to a $150 million market cap company. We raised over $40 million in one year. We went from twenty employees to fifty-five. Granted, many different factors influenced such significant growth, but I believe none of it would have happened without the cohesiveness we felt as an organization.

Rapid growth can have a downside, however, and for us, it meant recognizing who was not the right fit for the team. I did have to make some tough choices and let go of people who I did not believe were aligned with the organization or what we were trying to achieve. Our new approach to purpose gave me refined clarity, not only in terms of who were the right people playing for the team, but which position served each person the most.

None of this transparency came from any kind of annual performance review because I do not believe in the traditional formality of sitting down and evaluating an employee. In order to have real connection with the staff, I felt it was far more effective for department heads to do six check-ins with their direct reports a year, meetings I call "what-what's"—as in, what is working and what is not. The conversations are one-on-one, and the questions are posed to both parties. What-what's are casual. Participants can go for a walk, grab a coffee, whatever works for them, as long as they do it six times a year.

All of this contributed to our close-knit community, which only worked to strengthen and support each team member's purpose, and the results spoke for themselves. Extending the importance of coaching beyond myself and showing others in my orbit firsthand the benefits of it resulted in more accountability among each of us. This is a perfect example of what happens when you open doors for others—in this case, each team member was willing to go beyond their own formal roles and was motivated to take on more. Their connection to their purpose propelled them to go above and beyond. They put in extra hours, they completed important projects, and at the end of the day, everyone was more efficient and more effective.

Do your research before you hire a coach or a mentor. At the end of the day, you get to choose whom you give your money to, and if they can get you results, that is what ultimately matters. A good place to begin is by getting a referral from someone you trust in your industry and if your company has a program in place, start making connections there. Trust your intuition when you come across coaches with a too-good-to-be-true marketing story because they may not always be able to get you results in a way that feels good for you. A great coach will always have your best interests at heart, and you will know very quickly if you are both on the same team.

COACHES FOR THE WIN

One of the most invaluable things a coach can do is to show you your true potential, something that is not always completely clear for a lot of people. Back in my swimming days, I was a late bloomer, as I started competitive swimming at a much older age. Most kids start around age eight, but I was thirteen. Needless to say, I was the slowest and least skilled member of the team. But that did not matter to the coaches, who would soon become the most important influences of my life.

Coaches Andrew Currie and Rob Moretto saw what I was truly capable of. They knew my work ethic and they saw before I did how I would be able to shape my skills. There were two groups of swimmers: the age-group, made up of the younger swimmers, and the elite team, whose spots were reserved for the top competitors. With my lack of experience, I started with the age-group with Coach Rob, but Coach Andrew quickly recognized I probably would not connect with the younger kids and I would be missing out on the camaraderie that comes along with bonding with teammates my age. He encouraged me to start practicing with the elite group once a week. I eagerly did and joined them for dry-land training as well. It was a completely different training level beyond my competency, but Coach Andrew knew it would stretch my capability.

The workouts were rough, and I would not have been able to do them without his and the team's constant encouragement. For one exercise, we had to dive down and retrieve a twenty-five-pound block from the bottom of the twenty-foot-deep end. To me, it felt herculean, impossible. I was beyond nervous to even attempt it. I remember thinking there was no way I was going to succeed. As I dove down, the daunting nature of my task started to overwhelm me. But as I descended, I remembered every word my coach had ever said to me during training, encouraging me, reassuring me that I could do it. It was a voice I could not ignore. It got louder and louder as I reached the block. I propelled myself to the surface and emerged, brick in hand, with a newfound confidence in my abilities I otherwise never would have had. I felt an enormous sense of accomplishment around what my body was capable of, my own strength, and my mental endurance. After a few months, Coach Andrew and all of the elite swimmers officially asked me to join the team, a gesture that made me feel very welcome.

Without my coaches, I never would have known my potential, which is a key part of exploring your purpose. Coaches helps you understand your values, your skills, and your strengths while simultaneously ridding you of any limiting beliefs. They give you the supporting exercises needed to build confidence to reinforce your skills and motivate you to grow stronger.

Over the course of the year, I became a better swimmer. Thanks to their encouragement in a wildly supportive environment, I flourished, both in terms of my skill level and as a reliable teammate. I never would have asked to join the elite team on my own, and had it not been for the coaches' ability to see the best in me, I would have missed out on that entire experience.

The one thing the Gators Swim Club, where I swam for five years, excelled at was giving recognition as an incentive to improve. The coaches established a culture for recognizing people beyond their swimming capabilities, extending their philosophy of excellence into life beyond athletics. The year I joined the club, I was awarded Most Improved Swimmer and

Best Sportsmanship, and, for a beginner, this was uplifting. It gave me the platform to excel further because I was recognized for my hard work. For swimmers, it is expected that you will hit long stretches of plateaus in your career where it feels like improvement will be hard-won—the increments for gains are so small, sometimes by pure milliseconds. What set the Gators apart from other elite swim clubs was the recognition they gave to the swimmers, serving as fuel to achieve more, which is all too important for new athletes. We all need the positive reinforcement of a job well done as motivation to persevere.

Years later, I was lucky enough to become an age-group swim coach myself. I worked with the best, most respected coaches under the Canadian national swimming team. The entire experience instilled in me a deeper understanding of what true leadership looks like. It showed me the structure required to create and nurture Olympic athletes. It also taught me the power of individualizing programs to bring out the best in each of the athletes, with a few of them going on to become elite-level swimmers. In swim coaching, every day was a lesson in handling expectations—of myself, the coaching staff, my young swimmers, even the parents. Coaching has given me great confidence, and I carry the lessons I learned into every situation in my life.

THE TEAM

Now that you have a great coach at hand, you need to be open to finding the team that will propel you to victory. There is a huge relational component between purpose and interdependence. When most people are starting out in their careers, there is a natural tendency to focus solely on themselves. Their attention is on establishing themselves in their field and finding success. Their agenda is entirely about themselves, not about others. Recognizing the value in teaming with and relying on others requires a

BREAKAWAY

MUSCLE PLIABILITY

The idea of muscle pliability is related to the phenomenon known as hormesis, which is the process in a cell or organism that exhibits a modification in response to exposure to increasing amounts of a substance or condition. In strength training, if you stress the muscle, it regenerates and you have recovered and grown; stress it too much, however, and your recovery will take far longer, and you will not improve as much. There is a delicate balance that must be achieved.

Muscle pliability should be in any coach's wheelhouse. You and your brain will get to a certain point where you believe you cannot do more (remember the central governor theory?), but your coach will be able to tell you if that pain is well-founded or if it is purely fear. Your coach will be able to either push you further or encourage you to maintain your current level of fitness. If you decide to go it alone, you might experience a much slower growth because you might be too conservative or conversely, you might push yourself too hard and end up injured and quit. The coach's job is to help you find that equilibrium between a positive stress and a negative stress, the sweet spot where you can find progress without regression.

maturity most people only earn once they move past the individual satisfaction phase. The longer you stay in your career, the more you realize how working with others helps you identify what you are passionate about and inspire you to find new levels of happiness and fulfillment.

People who focus only on themselves will soon realize that it is hard to materialize bigger dreams without the assistance of others. It is extremely valuable to bounce ideas off trusted colleagues. When you lead an authentic, purpose-driven life, you recognize the importance of commitment and collaboration. You come to understand that executing your purpose requires a significant relational component. Coaching can help you understand that relationship faster because it helps you focus on the things that carry the most meaning to you, which inform how you can affect real change in the world.

A disclaimer: some coaches simply are not a good fit for you or your team. In my swimming days, I had a string of incredibly encouraging and effective coaches who were able to bring out the best in each athlete. We were like a family, and when one coach left, another person from inside the family would take over his role. In one case, when Coach Rob left the team, his younger brother, Paul, took over the reins. But one year, we had to bring in someone from outside the family, a man named Eugene, and he made his expectations abundantly clear.

"I am not Eugene," he said. "You can only call me Coach or Coach Eugene."

He was a disciplinarian, no question, and there were times I felt like we were pushed past our limits for his own gratification. He was known for these taxing workouts called negative splits. We had to run laps, increasing our speed with each one. If you slowed down, you would have to continue running until you hit your negative split. If you did ten minutes on the first round and nine on the second, you were expected to do at least 8:59 on the third. If you came in at 9:01, you had to do it again, no matter how spent your body was. This resulted in more than a few situations when people got pretty bent out of shape and walked off. I could see the intent behind the workout, but at what point does pushing someone to the point of giving up do more harm than good? Coaches must recognize when their athletes have given it their all and have nothing more to give.

Coaching is not about pushing you to the point of no return. It is about recognizing your unique skill set and knowing how to get the most out of that. Your coach should know who you are and what your capabilities are to help you execute your purpose to the best of your ability. Every person wants to find meaning in what they do, and part of the coach's job is to help you find that meaning. Together, you can work toward accelerating your path to excellence.

□

WORLD-CLASS CODA

- If you want to master anything, you must be open to finding a coach who will not only support your endeavours but also continuously challenge you to meet and exceed your goals. Egos aside, you have to fully accept the idea of being mediocre before becoming great.

- The perfect coach for you will be able to see your potential and know when to push you and when to hold back. Trust in the process because these are the times where you find out what you are made of, which can be very informative in hindsight.

- The role you play on a team is just as, if not more important, than flying solo. Never underestimate what can be achieved with others, and never hesitate to support those around you. The rewards that derive from true collaboration are infinite.

CATAPULT FORWARD

- What are your strengths and weaknesses? Are you able to pinpoint how they help or hinder you in life?

- How do you respond to stages of learning? Do you get frustrated when you hit a roadblock or are you invigorated by it? What are the instances when you give up and when do you persist?

- Can you think of an experience when you shared victory in a team setting? How did it differ from a time when you succeeded alone?

When you zero in on your paramount motivations for this life, you will find yourself on the path to greatness. In greatness lies the heartbeat of service. At your truest potential, your dedication to serve will reverberate, mobilize, and encourage the greater good.

PRINCIPLE 3

ON YOUR MARK, GET SET...

"Purpose is the essential element of you. It is the reason you are on this planet at this particular time in history. Your existence is wrapped up in the things you are here to fulfill."
— CHADWICK BOSEMAN

People living in the southern Japanese prefecture of Okinawa historically have been known for longevity and are more likely to reach the age of 100 than anyone else in the world. Perhaps it is their diet or maybe it is due to their environment. Regardless, research has shown that one of the predominant factors of their longevity is of a spiritual nature—people have "a reason for being," which they call "ikigai," the idea of having a purpose in life.[2] In Okinawa, there is no word that denotes "retirement." Ikigai imbues your entire life, unfolding and evolving as you live. In his book, *The Happiness Equation*, Neil Pasricha tells the story of a 102-year-old karate master whose ikigai is to carry forth his martial art; a 100-year-old fisherman, who catches fish three times a week for his family to eat as a gesture of provision; and a 102-year-old woman who embraces her ikigai in caring for her great-great-great-granddaughter.[3] Without a compelling answer to the question, "Why am I here?" we struggle to stay motivated. When our highest purpose is the elevation of our pleasure, comfort or status, it prevents us from growing and experiencing soul-level success.[4] [5]

Modern science and ancient wisdom seem to agree on the benefits of having a purpose in life. Purpose is what creates motivation, determination, and ultimately a feeling of happiness. Think of a sports team on a winning streak as each player on the team builds on their momentum, win after win. Everyone's purpose is not only to play their best game in every match, but also to elevate their personal game to maximize their chance of success. Purpose-driven people are much more likely to enjoy their work, be a leader, earn higher incomes, cope better with problems and stress, have

2 Marc Winn, "What is Your Ikigia?" The View Inside, May 14, 2014, http://theviewinside.me/what-is-your-ikigai (accessed on December 12, 2020).

3 Neil Pasricha, *The Happiness Equation: Want Nothing + Do Anything* (United States: G.P. Putnam's Sons, 2016).

4 Linda Wasmer Andrews, "How a Sense of Purpose in Life Improves Your Health," *Psychology Today*, July 14, 2017, https://www.psychologytoday.com/us/blog/minding-the-body/201707/how-sense-purpose-in-life-improves-your-health (accessed on December 10, 2020).

5 Alice G. Walton, "The Science of Giving Back: How Having a Purpose is Good for the Body and Brain," *Forbes*, July 10, 2017, https://www.forbes.com/sites/alicegwalton/2017/07/10/the-science-of-giving-back-how-having-a-purpose-is-good-for-body-and-brain/?sh=3dd00f9f6146 (accessed December 10, 2020).

better memory and brain function, sleep better, take better care of their health, play more sports, and avoid being depressed.[6]

To find your ikigai, you need to find good answers to simple questions. Do you want to enjoy life? Be really good at something? Make money? Help others? Instead of separating "work," "self," and "life," the notion of ikigai is to combine these seemingly conflicting goals in the right balance.[7]

If you are paid to do something you are good at, you have a profession. If you are paid to do something the world really needs, you have a vocation. If you love to do something the world really needs, you have a mission. If you love doing something you are good at, you have a passion.

Even when you find good answers to three of the four questions, you may still feel like you are missing something. For instance, when you are paid to do something you are good at and the world needs this service, you may still have a feeling of emptiness since you do not love what you are doing. Balance is only found at the intersection where your passions and talents converge with the things the world needs and is willing to pay for[8]. The clearer you are about your ikigai, the better you are able to channel your talents and time to make a meaningful mark on the world.

Every human being on this planet has purpose. Each one of us has something inside of us, driving us and propelling us forward every day, and

---

6 Ibid.

7 Chris Myers, "How to Find Your Ikigai and transform Your Outlook on Life and Business," *Forbes*, February 23, 2018, https://www.forbes.com/sites/chrismyers/2018/02/23/how-to-find-your-ikigai-and-transform-your-outlook-on-life-and-business/?sh=5f4352772ed4 (accessed December 10, 2020).

8 Marc Winn, "What is Your Ikigia?" The View Inside, May 14, 2014, http://theviewinside.me/what-is-your-ikigai (accessed on December 12, 2020).

understanding what that is and how to use it is perhaps one of the most important journeys we can embark on. The process of articulating your purpose and finding the courage to live it is the single most important developmental task you can undertake as a leader. Purpose can be a lot of different things to a lot of different people, but once you harness yours on an individual level, there is no limit to the positive impact you are able to have on your family, friends, community, and coworkers.

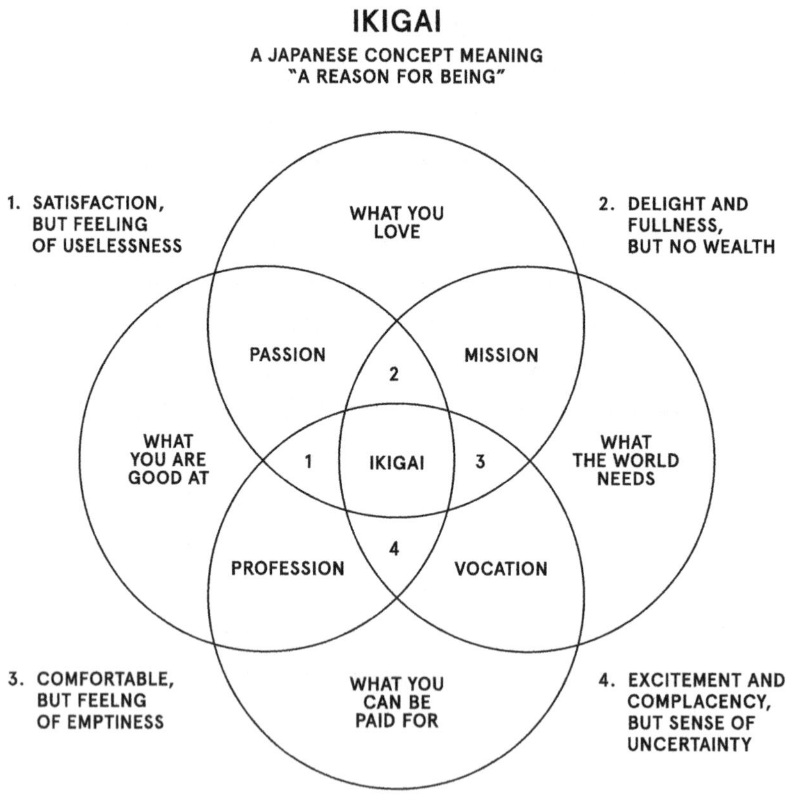

IKIGAI

A JAPANESE CONCEPT MEANING "A REASON FOR BEING"

1. SATISFACTION, BUT FEELING OF USELESSNESS
2. DELIGHT AND FULLNESS, BUT NO WEALTH
3. COMFORTABLE, BUT FEELNG OF EMPTINESS
4. EXCITEMENT AND COMPLACENCY, BUT SENSE OF UNCERTAINTY

WHAT YOU LOVE
PASSION
MISSION
2
WHAT YOU ARE GOOD AT
1
IKIGAI
3
WHAT THE WORLD NEEDS
4
PROFESSION
VOCATION
WHAT YOU CAN BE PAID FOR

Ikigai[9]

9 Adapted from Héctor Garcis et al., *Ikigai: The Japanese Secret to a Long and Happy Life*. Penguin Books, 2017.

To understand purpose, it is helpful to first recognize what it is not. It is not a target you are trying to hit, nor is it a jargon-filled catchall. It does not have to sound amazing, aspirational or even altruistic. It simply must be the thing that consistently fuels you with the energy you need to make the changes you want to make for yourself and others.

Purpose is something you cannot help being. It is not the job you do, but *how* you do your job and why. Your unique purpose is rooted in the strengths and passions you bring to the table, no matter where you are seated. Yet finding your leadership purpose is not easy. If it were, we would all know exactly why we are here and would be living that purpose every minute of every day. We are constantly bombarded by powerful messages from parents, bosses, management, gurus, advertisers, celebrities, and social media influencers about who we should be and how to lead. It is easy to emulate someone else's purpose instead of being guided by your own. To figure out your purpose is not easy, but once you do, everything begins to flow naturally.

Purpose is directly linked to your ability to demonstrate authenticity with consistency. Authenticity is honouring your values and integrity. It is what gives you the ability to react or respond in any situation because you know who you are and what you will or will not stand for. By the same token, your environment and experiences help to establish resilience because your experiences are invariably linked to your values—how you react to what happens to you, success or failure, is dependent on your character and the things in which you believe. Therefore, your authenticity gives you the structure necessary to continue to lead with purpose. When you are being authentic, others recognize this. It inspires others to put their own purpose into action and truly live the lives they are meant to live.

LEADING WITH PURPOSE

Honouring your values is key to leading with purpose. When you lose a sense of your values, it is easy to become misdirected. Your decision-making skills can become compromised, eroding your goals, leaving you with a loss of purpose. Before long, you begin to question why you are investing so much of your time and energy into the work you are doing.

Strong values are inextricably linked to your purpose. To protect yourself from the pitfalls of losing sight of your values, it helps to identify the factors that could cause you to compromise them. Is it too much pressure from stakeholders? An unexpected downturn or hurdle? Complacency? Pressure from colleagues? Awareness around the things that can derail your path is a key part of your Corporate Athlete strategy and for keeping yourself anchored in times where outside forces cause you to question your objectives. Holding firm to your own set of values allows you to navigate the corporate world with more confidence. A threat to your values is a threat to your purpose.

Purpose feels like a grand word, but in the end, it is about helping yourself and others see your individual impact and developing a deeper understanding of why we love what we do. If you keep that in mind and take a personal and an authentic approach, you are likely to find success.

As a world-class Corporate Athlete and leader, helping others feel a sense of purpose is also a very powerful tool. Over the years, I have tried and succeeded in lighting the purpose fire for different employees and teams I have worked with. There have also been times where it has been challenging. In my last fifteen years of experience in executive positions, I have come to the conclusion that it is extremely difficult to instill purpose in others. It takes more than motivational talks, speeches or mission statements to spread purpose. In fact, if overblown or insincere, those methods can backfire, triggering cynical reactions. Actions speak louder than words and creating opportunities for participation and collaboration allow people to tap into their purpose when they otherwise might not have the chance to do so. For example, I have used my passion for athletics to bridge my purpose and establish community participation through group rides, runs, and other activities. Connecting something I love doing with the workplace has fostered involvement and interest with my colleagues that would not have existed if we only focused on the work in front of us.

Working with Judy, I have encouraged over fifty different executives from various companies to understand their personal profile and performance assessments. We have introduced them to the immersive Brew Creek experience to help each person identify with their individual purpose. From there, we have spent time to ensure their jobs are aligned with their purpose, making changes to their responsibilities, milestones, and sometimes even job titles. What has been unique about the immersive experience at Brew Creek was the personal attention given to each participant, from the moment you stepped foot on the grounds to the moment you left. Purpose needs to be personal, and each of the workshop sessions was crafted to elicit an emotional reaction, the personalization allowing purpose to be felt. You cannot just talk about purpose. When it comes to a job, it does not matter what line of work you are in, everyone needs to be able to see and measure the impact of their work to feel connected. When the different executives I have worked with see the cause and effect between their input and their team's progress, understand the impact a patient is getting from a potential new drug, or experience firsthand how their role is necessary to an unmet need, they have all felt a sense of purpose. The challenge comes when people do not see or believe that what you are doing is meaningful because the payoff may not be immediate, especially when it comes to linking urgency to products that may not be commercialized for years. Authenticity toward your purpose can make all the difference. If your attempts at creating purpose do not fall in line with your other leadership behaviours, employees will view your tactics as manipulative rather than genuine. Once you have made it personal and authentic, you cannot do it just once—you need to make it a habit. This is leading with purpose.

Imagine a sports team: If the coach is effective in communication of the team's purpose and leads with this in mind, then naturally, all members of the team are working toward a common goal, even though each player may have a different role. There is a powerful effect when a team of individuals who know their purpose come together. For example, Wayne Gretzky did not win the Stanley Cup alone. The coach guided the team with the sole purpose of winning the cup, and the team worked hard together. Doc Rivers, who coached the 2008 Boston Celtics to the NBA Eastern Conference championship, had a mantra based on this very idea: ubuntu. Ubuntu is a concept that revolves around the ideas of community and collective success. That mantra created a shared mindset among team members, who knew that they won together, they lost together, and no one alone was responsible for any of it—and it clearly worked.

Teammates empower one another. If you have ever been on a sports team, you know how important it is to be present and how the rest of your team is relying on your skills. When the entire team operates within their personal ikigai, success is all but guaranteed. Everyone shows up to perform to the best of their ability because they are aligned with their purpose, demonstrating accountability to one another, the coach, and the team. Here, leading with purpose does not necessarily fall to one person, nor should it—in a team, everyone is a leader at different times, bringing different strengths to the forefront in order to create the optimal environment for success.

BREAKAWAY

WHERE IT ALL BEGAN

When I was twenty-six, my friend Maggie casually mentioned in conversation that she was signing up for the Carlsbad Half-Marathon. When she told me she was running, she challenged me to run as well, but because she told me with such short notice, I only had twelve days before the race in order to prepare. Keep in mind that, at this point, I had never done anything more than a six-mile run in my life. Little did I know, this would be the race that would get me hooked on long-distance running.

I emailed my oldest and best friend, Tu, who had prior experience as a personal trainer, asking him to design a training plan because I did not have any other coach to rely on. He thought I was crazy for attempting to do this in less than two weeks, but he relented. I did everything according to that training plan. I am sure it was not what a running coach would have recommended, but because he had some authority and I needed structure, I relied on it. I ran the half-marathon and came out relatively unscathed.

Right after I finished the race, my first thought was that I did not want to run ever again, even though the race was perfectly fine; I was done. But inexplicably, a few minutes later, I wondered quietly when the next one was taking place. I found out and signed up. I was bitten by the bug.

After that, I continued to compete in many half-marathons in order to focus on my speed. There was a marked difference between my first race, the Carlsbad Half-Marathon to one of my later races, the

California International Marathon, where I felt incredible running because I was running with purpose. The growth I saw from one to the other was remarkable.

Tu designed the first running plan for me, believing in my outlandish ideas, without knowing he was going to give me the nudge necessary to become very passionate about running. When you surround yourself with people who believe in your abilities, despite data, history, and common sense, it can lead to uncharted territory.

If I had not tried running a half-marathon, I would most definitely have had a very different career since running has helped me to continuously evolve, survive, and thrive in the corporate world. It has given me time to think about my purpose and what matters since this concept is laden with so many meanings. It created the very idea for this book, as the core insight allowed me to make a meaningful connection between athletic metrics and business metrics. When I witnessed the value of running and the rituals that sport establishes and realized the positive effect it had on my life, I became a more disciplined, focused person. The dramatic influence running has had in my life undoubtedly is positively and directly correlated to the overall satisfaction in my career. At the same time, running has helped to break down walls and barriers within organizations and also outside them—hardly anyone turns down an invite to a run. I have translated this sense of purpose from running to business organizations to stress the impact that it has upon people's ability to grow, innovate, and transform. Purpose matters, both in and out of the office, and it can only have positive effects on an individual's and a corporation's performance in the present as well as the future.

FAST TRACK

Many leaders and changemakers do not take the time to fully explore their purpose until they are deep into their careers, which can result in missteps, miscommunication, and missed opportunities. The intention of this principle is to get you on the fast track to finding your own purpose by challenging you to dig deep to discover what you are made of. This will allow you to have an even more meaningful, significant impact on the world while also creating a framework to check in and evolve/adapt your purpose. It will help you realize what excites you, what motivates you, and what unique contributions you can make.

This may not come easily for everyone. Some people possess a natural knack for introspection and reflection. Others might find the experience uncomfortable and anxiety-provoking. No matter how you are feeling, I encourage you to explore the questions posed in this section with openness and honesty. To effectively unearth the most helpful answers:

1. Take time to think and write down an essay-style answer to each question. Feel free to come back to it and edit as you go along.

2. Go to a new/novel location to get your brain excited (preferably a location that connects you to your purpose).

3. Forget the written answer for a week, then reread and edit it.

4. Tell someone about it. Discuss the ideas you have written down or new ideas that come to mind as you chat.

When you are ready to dive in, pick the questions that resonate the most with you, put this book down and get started. Then revisit this list as often as you feel is necessary. By no means is this meant to be an exhaustive list, but it serves as inspiration for you to think creatively about the things that are most important to you.

SELECTED QUESTIONS FOR DEFINING PURPOSE

- What activities would infuse my life with more joy and meaning?
- What should I be doing with my time and my abilities that would be helpful and make a difference in my life for those I love and potentially for others?
- What direction(s) should I be pursuing that will feel better to me than this boring, meaningless work I am engaged in now that leaves me feeling empty?
- Do I feel I am wasting my time in work that is just a paycheck and nothing more? How can I stop that feeling?
- If I was not working, what would I enjoy?
- What can I do in my time off from work that will help me feel that my life matters more?
- What can I do with my time that is important, meaningful or enjoyable?
- What is true about me today that was also true about my younger self?
- How can I show vulnerability and grow?
- How am I going to save the world? What solutions can I offer to some of the challenges in this world?
- If I had to leave the house all day, every day, where would I go and what would I do?
- If I knew I had limited time left in my life, what would I do and how would I want to be remembered?
- If I could improve the world in one way, what would it be?
- What kind of problems do I love solving that others might dread?

- What underdeveloped skills do I have?
- What kind of people love listening to me and what kind of people do not?

PURPOSE IN ACTION

After you have written down your answers, it is time to articulate your purpose in a statement. The act of mining your life should have revealed some common threads and major themes. By now, you should have identified your strengths, values, and things you love to do and should have some ideas about how everything is working together to ignite a passion inside you. Now, it is time to write your Purpose-to-Impact Plan to illustrate how you can use your unique leadership purpose to envision big-picture aspirations, then reverse engineer to set more specific goals.

Purpose-to-Impact

1 Create a Purpose Statement

Read through your answers. What are some of the common words? Use them to craft your statement. It can help to read over a few examples to land on the words that work best for you. For example, Sir Richard Branson has said his purpose is, "To have fun in [my] journey through life and learn from [my] mistakes." Oprah Winfrey's is, "To be a teacher. And to be known for inspiring my students to be more than they thought they could be." You might recall my purpose is, "leading people and communities to pioneer indelible impact." Each word in that statement carries significance to me. What words speak most to you?

2 Write an Explanation

I love to work on new business ventures. In my late teens, I was exposed to different business experiences, having come from an entrepreneurial family. I also lived through the advent of the Internet and the Dot-Com era, a period of massive growth and innovation. In my early twenties, I witnessed the tremendous importance of the healthcare sector by gaining experience at a healthcare venture capital firm. Now, working on business opportunities in the life science area is my passion and I am driven by the opportunity to bring new technology forward for different diseases that have an unmet medical need. How can you justify your purpose?

3 Set One-Year Goals

I want to foster the growth of a business while remaining consistent with reaching the different inflection points for a technology. I will support the R&D team operationally and assist in ensuring the resources are available to carry out different research objectives. If there are failures along the way, I strive to embrace a 'fail-fast and fail-forward' mindset to learn and adapt from these experiences. What long-term goals will keep your purpose on track?

4

Set Two-Year Goals

I aim to qualify the technology or product through the different regulatory steps to advance development toward approval. To do so, I will ensure that each goal and objective is Specific, Measurable, Attainable, Relevant, and Time-based (SMART). I will ensure adequate funding is in place to carry out the clinical related research for the technology or product. What goals will strengthen your ability to live your purpose?

5

Set Three-to-Five Year Goals

I can start by targeting shorter goals and milestones for the product development. I will work with the team to seek input from researchers and key opinion leaders to ensure the product merits a commercial development plan. I will begin to map out a detailed plan and demonstrate differentiation in the marketplace. How can you begin to think past your own finish line?

6 Map Out Critical Next Steps

Think practically about how you will execute this plan. In my case, I will examine a detailed manufacturing, clinical operations, and development plan in order to secure adequate resources to execute the critical steps and objectives. I am going to think about every aspect of the operational plan all the way down to which day the data will be published on our product. What actions do you need to take to ensure success?

7 Examine Key Relationships

In order for me and my team to devote the time and energy needed to execute the development plan, I will have alignment with key stakeholders, such as the board of directors of the company, management, as well as key investors. The alignment will encompass buy-in on the essential milestones for the development plan. My key stakeholders must be in agreement with it. It cannot impact any other aspect of my purpose in any negative way. If it does, I have to revisit how it is really aligned with my purpose. What impact will your actions have on the people who matter most to you?

BREAKAWAY

WHAT ARE YOU DOING?

We all occasionally do things without thinking, and this is where we set ourselves up for failure. I used to believe the pinnacle of completing an endurance event was to arrive at the finish line completely breathless, totally annihilated, and on the verge of collapse, proudly knowing that in my heart, I finished the race to the best of my ability. The funny thing is, sometimes you end up with an IV in your arm instead of a medal around your neck and realize the whole experience was your own undoing.

In April 2006, I competed in the La Jolla Half-Marathon and was on track to finish a 1:35 half-marathon. I desperately wanted to prove to my brother that I was faster and that was my only goal—not running a personal best or enjoying the run with him, but simply winning. I was also running on behalf of my company and felt like I had to prove myself because I had encouraged everyone to participate. This all added up to a great deal of unnecessary pressure I placed on myself.

In an alternate world, if I finished the race well ahead of my brother, I would have emerged victorious, and the story would have ended there. I would have been thrilled, even if I was never to take another step on a half-marathon course again. But the reality was so much more layered because when you just have your eye on the prize and nothing else, little good can come from it.

I came out of the starting line hot, maintaining an unsustainable seven-mile-per-minute pace during a scorching-hot San Diego day. There was an incredibly difficult 400-foot climb up Torrey Pines in the middle of the marathon and I was not taking any hydration because my narrowed focus kept me running fast and hard, purely lasered in on winning the race. Around mile ten, I started to feel hazy; by mile twelve, it was the beginning of the end. Unsurprisingly, I fainted one mile short of the finish line. My brain and body were completely disconnected. I was told after the fact that I was begging people to give me their shirt in complete delirium. I had to be rushed to the aid tent via ambulance—it was all very non-victorious and completely anticlimactic.

This race was meant to be the second out of three that would give me the Triple Crown in San Diego, a title awarded to people who had completed all three marathons in the city. Still obsessed with completing the half-marathon to get that notch on my Triple Crown belt and delusional due to the dehydration, I desperately tried to convince the paramedics to drive me to the end of the race so I could toss my shoe with the race microchip over the finish line, just to say that I finished it. I was clearly out of my mind.

As with most mindless decisions we make, this half-marathon and what I wanted out of it were all misaligned. I wanted an outcome, but it was not in keeping with who I really was and why I love to run in the first place. Sure, friendly competition with my brother is fun, but when I compete now, I do it for many other reasons other than just winning. Winning simply cannot be the be-all and end-all. In the end, it is important to take stock of why we do the things we do and recognize it is a privilege to even be performing at this level. When we are clear with our intentions, we tend to avoid situations such as this one, where the need for victory keeps our blinders on and we miss what the race could truly be about.

DIVING DEEPER

Your purpose should be a source of inspiration for yourself. It should always be there to motivate you and hold you accountable. Even if you feel comfortable with where you are at in your life, knowing and understanding your purpose is critical. Becoming intentional about your purpose will lead to more transparency, more positive behaviours, and ultimately, a more significant impact in this world.

REVISITING PURPOSE

As you continue to develop your purpose, it is key to remember it is not something you find, but something you continue to build on over the duration of your life. Nor is it just one thing, as no one sentence can encompass all your meaning and everything you are capable of contributing to the world.

To stay aligned with your purpose as it continues to evolve, I encourage you to revisit it as often as you feel necessary. The more often you do, the more familiar you will become with it and the more you can refine it. It is especially helpful to revisit your purpose when you are facing a challenge, as it can remind you of the process you went through to define it or even give you the answers you may be looking for.

ORGANIZATIONAL PURPOSE

While we spend a lot of time trying to find individual purpose, it is also important to remember purpose is much bigger than yourself and consider how it impacts your professional life. Organizational purpose is one of the key qualities for having a lasting company. Many companies (generally start-ups) find purpose early because they want to solve a specific problem, or they see a solution.

As an organization grows and evolves, so does its purpose. It can evolve negatively when people lose their individual purpose or become focused on things outside of the organizational purpose. Therefore, companies also must revisit and reevaluate their purpose to remain consistent on their shared goals.

ALTRUISM, SERVANT LEADERSHIP, AND IMPACT ON SOCIETY

People often think their purpose should be linked to a specific cause or have a significant social aspect. I do not believe this to be true. It is more important that your contributions have a positive impact on yourself, on your workplace, and on the people around you. It is great if your purpose is to save the whales, protect the environment or end hunger. But if it is not, you do not have to put unnecessary pressure on yourself to focus on something altruistic in nature.

Furthermore, purpose is influenced by many factors, including cultural norms and expectations. As you are shaping your purpose statement, consider your environment and unique circumstances, allowing your purpose to support them.

That being said, I encourage you to look for opportunities to use your purpose and your role as a leader to serve others, as opening the doors for others can create a wave of positivity that will affect every aspect of your life.

LIVING YOUR PURPOSE

By now, you should have a solid sense of how to fast-track your purpose. You should feel resolute and ready to put it into action. Remember that the work is not complete. Why are some individuals and organizations better able to strategically integrate and capitalize on purpose than others? How are they doing so? It is clear that maintaining intentionality around purpose is a never-ending process and if there is insufficient leadership commitment, it will most definitely lead to misaligned performance metrics. This principle has hopefully provided you with a starting off point, and the subsequent concepts you will learn in the book will help you continue to develop and grow your purpose as you learn more about yourself. All of this plays an important role to drive the changes you want to see in yourself and in your career.

As with many things in life, better questions lead to better answers and therefore, a better understanding of purpose. After all, what creates the greatest runners, entrepreneurs, leaders, and companies? Each operates from a slightly different set of assumptions about the world, their industry, and what can or cannot be done. That individual perspective allows them to create great value and have significant impact. They all have a unique leadership purpose. To be a truly effective leader, you must do the same. There is no time like the present—clarify your purpose and put it to work.

□

WORLD-CLASS CODA

- Purpose is the compass that will consistently guide you in every personal and professional decision you make. When you are clear in your values and the reasons why you do the things you do, it will resonate in your actions. When your purpose and action are in alignment, it will be easy to know when you are on track.

- Goal setting is critical to success—long term or short term, defining the things you want is paramount to creating a map toward achieving them. A purpose-to-impact plan is deliberate, specific, and transparent. It will keep you accountable, turning your dreams into reality.

- Beyond your own personal achievements, using your purpose for the greater good is not only admirable, but also necessary in today's climate. Figuring out ways you can help solve critical issues within your community is a catalyst for global change.

CATAPULT FORWARD

- Is there a quality in the adult version of you that you had as a child? Can you connect this to your purpose? Was your purpose clear back then to you as a child or has it changed?

- What is one current short-term and long-term goal you want to achieve? Is there one thing you can do today to help get you there?

- Is there an issue in your community you think can be helped or solved by your particular skill set? How and what can you do to make this happen?

TOGETHER STRONGER

SECTION TWO

Rote learning is unappealing. The mechanical day-to-day repetitions can feel like they serve no purpose. All the practicing, to what end? There is, however, an exquisiteness in the rituals you undertake, the small victories that add up over time to greatness. All the blood, sweat, and tears are mired in the everyday work no one sees. While the dress rehearsals are never as polished as the live performance, one is born out of the other; what you put into it is simply what you get out of it.

PRINCIPLE 4

FINDING YOUR STRIDE

"Real change, enduring change, happens one step at a time."

—RUTH BADER GINSBURG

It took nearly drowning before I learned how to swim—and that is not just a metaphor for the many career-ending mistakes an entrepreneur can make in their early days. I took my first swimming lesson very shortly after a near-death experience in a relative's pool. I was nine years old, and my parents had brought me to my aunt and uncle's condominium outside Vancouver for a visit. The building had a beautiful, hexagonal-shaped community pool that was, in my memory, enormous. Despite having no aquatic expertise, I wanted to be in that pool so badly that when the adults were off doing their thing, I wandered over to the pool and stepped into the very short, shallow end.

After a few small steps, there was nothing beneath my feet. The floor dipped down into the deep end and I was left thrashing furiously to keep myself afloat. Panic surged through me as I began to realize I might not make it. I must have caused enough of a commotion that someone—to this day, I still cannot remember who—shoved a pole into the pool within my grasp and pulled me to safety.

My relatives ran over and bundled me up into a large towel, angry at my carelessness, but mostly relieved because they knew I had learned a big lesson that day. The moment my dad knew I was out of harm's way, he had exactly one thing to say: "You are taking swimming lessons."

Whether it was because of my dad's insistence or because of my own desire to never feel out of control in the pool again, I became religious about swimming lessons and, later on, competitive swimming. I have vivid memories of my mother driving me to the pool on Tuesday, Thursday, and Saturday mornings before sunrise, rain or shine, school day or weekend. By the time I was twelve, I was a junior lifeguard, and by thirteen, I began swimming competitively. At this point, I was mildly arrogant—a predictable cocktail of teenage cockiness mixed with the self-assuredness from excelling at the sport. I thought I knew everything about swimming; I had taken so many lessons, I could teach them myself. But I was in for a rude awakening. I soon learned that swimming lessons and competitive swimming are like apples and oranges—comparing the two is akin to taking a leisurely stroll and successfully running a 5,000-metre race.

In the summer of 1993, I was enrolled in a swim camp that met for an hour every morning. I remember thinking, "This will be easy; I got this." It only took one drill for me to realize that I indeed did not have it. I could not wrap my head around the amount of endurance the sport required. But I stuck with it, worked hard, and by the end of the camp, my performance had greatly improved. So, I joined another camp. I kept swimming at least an hour each day. After a month, I was stronger, faster, and had a great deal more respect for the sport.

Since that summer, I have participated in a number of swim training camps, which are short, very focused experiences. In sports, when you are looking for millisecond improvements, it is important to focus on something different in every session to realize these gains. Sometimes, it may exist in the form of swimming stroke techniques, pacing, breathing or counting drills. My swim coaches always underscored how repetition is the highest form of learning. Now that I have been out of the water as a competitive swimmer for some time, I have a better appreciation for how this practice of repetition has also helped me as a person, leader, and Corporate Athlete. In the corporate world, off-site events are concentrated but offer focused time, team-building opportunities, and other personal development activities most Corporate Athletes do not frequently experience. These events are like training camps for the professional athlete—they do not last long, but they prepare you for the rest of the year. It is important that you make the best use of your time during these events in order to maximize and build on the skills you have in your repertoire, setting yourself up for the best chance of success throughout the entire span of your career.

TIME AND EXPERIENCE

One of the greatest myths is that experience at work is measured in units of time. But time is not an accurate measure of experience. For example, if you are hiring and interview two candidates for an executive position, one with fifteen years of experience and the other with twenty years of experience, your automatic assumption would be the latter candidate has greater experience. Yes, to a degree if we are purely speaking about the numerical value of time. But what if the person with twenty years of experience worked in a mostly automated role, while the other person worked in an accelerated, learning-oriented way, catalyzing their work and assimilating it into technically more quantifiable years of experience? The number of years a person spends at a particular workplace is a measure of just that—how much time they have spent at work. It does not measure how powerful their learning was during that period and it is not a measure of the experience they have built in the process.

Entrepreneurs tend to quantify experience by time and time only, and I believe it is important not only to reveal this blind spot, but also to address it. In comparison, athletes also invest time in their craft, but it is not the only thing that makes them great at their sport. They are constantly working toward smaller goals that build incrementally toward bigger ones. Year after year in swimming, I made many small sacrifices, going to sleep early on Fridays instead of hanging out with my friends, so I could be up and rested early for practice on Saturday mornings. This focus and discipline created resilience because I understood what it took to achieve the bigger goals that I set for myself. Today, I use the same disciplined approach to pursue the things that are meaningful to me. Athletes spend inordinate amounts of time reflecting on the things they want to achieve, sometimes on a daily basis, depending on how the training is progressing; entrepreneurs lack this regular introspection, which, I think, is sorely needed. How do we encourage entrepreneurs to reflect often on said experiences in order to improve their performance in the workplace?

I had begun my swimming career late. Most people start far younger than I had. But that only meant I had some catching up to do and this made me work even harder. In the end, the condensed experience helped shape who I would become as an entrepreneur and it encouraged my love of the sport. In my youth, I trained up to twenty-two hours a week between swimming, weight training, and running, all the while learning what I was capable of, what I could tolerate, and what improvements I could make as I perfected the craft. The sheer volume of physical work expected of me during training instilled in me a fierce discipline. The act of swimming consistently taught me the importance of having a "feel of the water"—when you experience the catch of the water and the sensation of acceleration, swimming feels effortless. It is similar to a runner refining their stride by tweaking the angle of their ribcage over the pelvis or a cyclist perfecting their pedal stroke by finding the optimal hip-knee-ankle alignment during each crank revolution. With focused practice, it becomes second nature and you can easily sense when that feeling is off. This is no different in business—you know when you are in the zone.

With time, I was slowly becoming more confident and perhaps, even more importantly, I was beginning to understand the value of being a part of a community. My parents and I had formed relationships with other swimmer families as well as the coaches and these people became an extension of our own family. Everyone supported one another. The adults would constantly reinforce our leadership qualities, saying things like, "Oh, Punit? He is definitely going to accomplish big things." As kids, we saw their example and became incredibly supportive of one another as teammates and peers. I was easily one of the worst swimmers on the team, but I was the biggest cheerleader at the swim meets.

Swimming deepened my own work ethic, creating a firm foundation for my future as an entrepreneur. Without it, I would have never had the discipline or the ability to look outside myself, which is absolutely required for success in the business world. A firm foundation means no cracks, and as an entrepreneur, certain behaviours can guarantee shaky structural integrity. Just as a defeatist attitude in athletes guarantees losing, self-doubt in entrepreneurs ensures failure. Athletics taught me that anything is possible if I just keep trying to push myself, especially when my goals seem impossible. Swimming created a structure and a source of resilience to train up to ten to twelve hours a week in a multidisciplinary way, even well into my forties. Currently, I swim with the Solana Beach Swim Masters Club in San Diego, and I still feel the familiar urge to compete in a swim meet. This discipline and dedication crossed over and has served me well in business, allowing me to foster resilience in my team and maximize the work environment to make the best use of time and energy. I could say the same for my teammates, many of whom went on to become very successful in their careers and prominent contributors to their respective fields. It also helped me find my groove because I fully understood who I was, what I wanted, and how I fit into the community. All of that reinforced my confidence, knowing I needed to perform at my best. The distinction between athletics and Corporate Athlete success is undeniable, and it is now easy to see why.

TRAINING MAXIMIZATION

Athletes maximize their training not only through time devoted toward the sport, but also through quantifiable measurements that reflect consistent improvement, even after they have found their stride. Corporate Athletes need to adopt a similar measure to regularly chart their progress. In both arenas, distraction, fatigue, and excessive work demands undermine your ability to focus and improve your performance against a target. Therefore, before achieving high performance in training and living, you need to recognize the things that hold the greatest value in your life. For you to be truly productive and facilitate your purpose, you need a healthy balance of both consistent performance review (whether it is from self-review or a coaching figure) and reaffirmation of personal discipline, which may include quiet time or concentrated work time to reinforce positive habits. This will require setting aside devoted time to make a schedule or structure that allows you to plan for success. As a world-class Corporate Athlete, the time you put into an activity is linked to meaningful experiences that develop the anything-is-possible resilient mindset. Different elements in work, self, and life are critical to your success—organizing your life to address the gaps is key to a balanced, purposeful approach to training maximization. What if the work that I put in as a competitive athlete and as a Corporate Athlete produces results beyond my wildest imagination?

Ironman distance (half or full) training is a beast to conquer. When I sign up for one, I work with my coach to draft a specific plan as preparation with the understanding that depending on various circumstances, I will have to make modifications along the way. The first thing an athlete does is to set a goal for the activity. When I made the decision to compete in my first Ironman, I wanted to train for it without compromising my commitments to work-self-life and to complete it without any injuries. Aside from a few bloody blisters, I met my target. My wife, Nina, and I have a running joke: Which is harder, training for an Ironman or having a child? (We plead the Fifth on this one.) Similarly, if I have a corporate goal, I work with my team to create a plan to achieve it, knowing I or the designated team lead will likely come up with some workarounds and solutions to get there. Athletics taught me that sometimes, you must be willing to change your goal altogether. There have been several times when I have signed up for races and ended up not competing due to scheduling conflicts or injuries. But those experiences did not take me down and they certainly did not mean I had no plans to give it another try when the time was right.

As is the case for any race, shortcuts are not acceptable in the world of entrepreneurship. Imagine training for an Ironman without a plan, flying by the seat of your pants? As a Corporate Athlete, it would also be a futile effort to maximize your individual or corporate results without a plan. If anything, you will be required to work even harder than you might anticipate in order to become world class. Think of it as super compensation, a weight-lifting principle based on the idea that when you do the heaviest lifting regularly, lifting anything lighter is that much easier. I have run long distances at night or before sunrise, sometimes in extreme temperatures, to prepare myself for the different elements I would encounter during the running portion of Ironman. I needed to know what that experience felt like, so I went beyond my routine training to make sure I was ready to take on the extra challenge. When I raced my first four-kilometre swim, I trained by aiming for six kilometres to make sure the shorter race was well within my ability. When preparing for a marathon, I always run extra mileage so when I get to mile twenty-five, I am certain I can get through the last mile. During training, the inner voice that reminds me to go one more mile, round, set or repetition is satisfying because it gives me reassurance that during competition, when the body and mind are battling each other, I have the will to keep going.

Psychologist and positivity expert Martin Seligman explored the concept of building resilience through a program called Comprehensive Soldier Fitness (CSF), based on the idea that the preparation of a soldier is not just about the physical aspect, but the mental as well[10]. CSF is a form

10 Martin E.P. Seligman and Raymond D. Fowler, "Comprehensive Soldier Fitness and the Future of Psychology," *American Psychologist*, 2011, 66(1), 82-86.

of training that teaches leaders how to embrace resilience by building mental stamina, signature strengths, and strong relationships. I believe the same approach can be applied to the idea of corporate fitness because every Corporate Athlete needs to have a thorough understanding of his or her capacity for resilience. Your foundation must encompass resilience in order to withstand stress or adversity, no matter what circumstances you may find yourself in. When you have a strong foundation in place, you can view setbacks as a chance to pause, analyze what went wrong and why, and move forward smarter and stronger than before. No athlete shows up to any competition expecting to lose. No matter the goal, you are only ever going to achieve the best outcome if you approach every situation with confidence.

If you have a real desire for change, you must have clarity in the things you want to achieve, in business and in sports, and if you lay out the roadmap, you begin to mark the paths that will take you where you want to go. Beyond this, if you want to ensure success in your undertakings, the most important thing you can do is prepare, overprepare, and have contingency plans at the ready in order to weather any storm. This creates resilience, and resilience is required for success. With a clear game plan tucked into your back pocket and with the unwavering support of your team, you can overcome any challenge with ease because the firm foundation that you have created will be more than enough to carry you through.

CORPORATE ATHLETE HIGH-PERFORMANCE TRAPEZIUM

Sustaining peak performance amid the increasing pressure you are under as an entrepreneur can seem daunting. With the likelihood that your career is going to be a lifelong journey that will evolve as you grow, you need to maintain your energy and enthusiasm while keeping yourself motivated and focused on your goals. A professional athlete understands the importance of improving themselves in all aspects of their life because competition at the highest levels requires more than just pure dedication to their sport. It is only those who truly master this who go on to become great, having long, successful careers.

Personally, I rely on an adapted version of the High-Performance Pyramid, a concept originally developed by Jim Loehr and Tony Schwartz. Their pyramid is comprised of four layers: physical, emotional, mental, and spiritual capacity.[11]

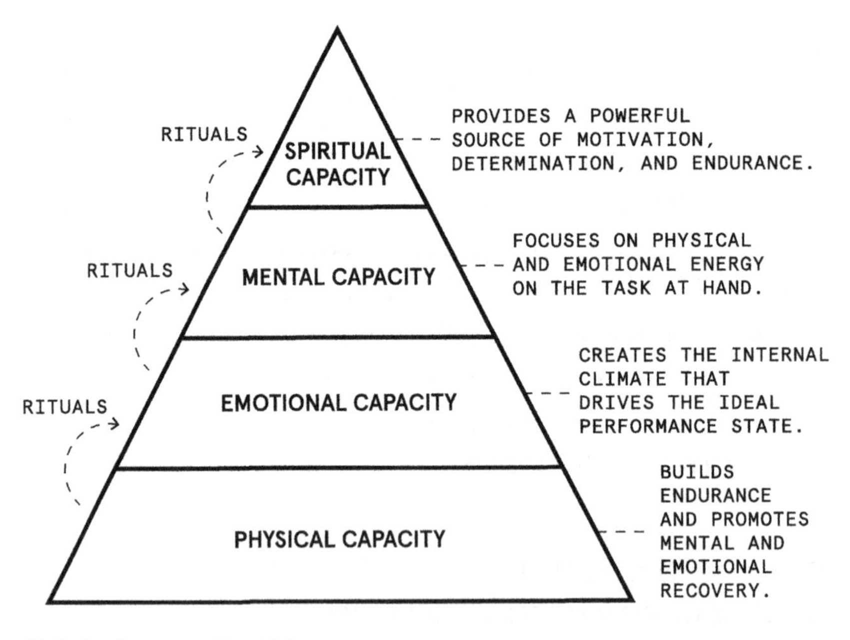

High Performance Pyramid

11 Jim Loehr and Tony Schwartz, "The Making of a Corporate Athlete," *Harvard Business Review*, 2001, 79, 120-128.

In my interpretation, the pyramid changes its shape and becomes a trapezium.

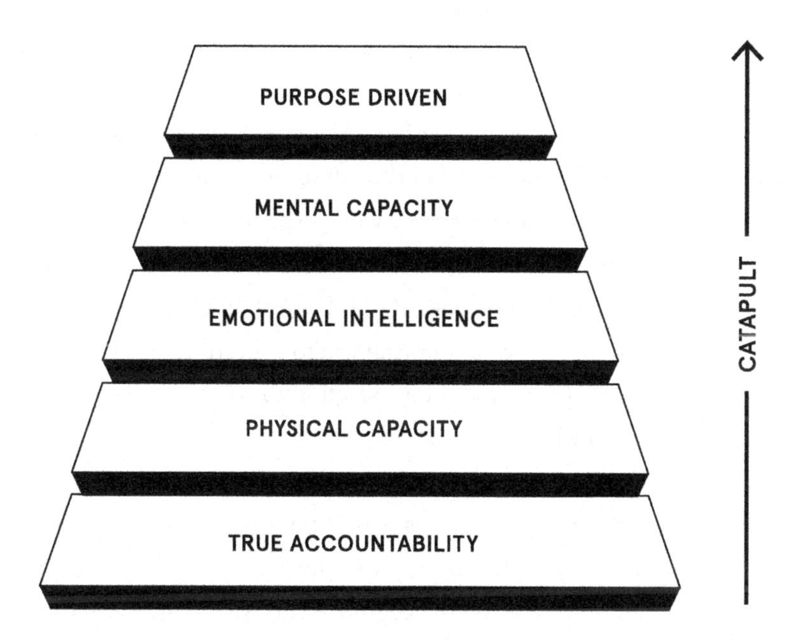

Corporate Athlete High-Performance Trapezium (CAHPT)

Here, the base layer of the Corporate Athlete High-Performance Trapezium begins with true accountability. As you already know, accountability, including your ability to trust yourself and commit to your values, is the bedrock of everything. The second layer is physical capacity, as building endurance and stamina help to promote all other aspects of the graphic. The third layer is emotional intelligence, referring to the establishment of a positive mindset. The fourth layer is mental capacity, comprising mental stamina and the ability to focus. Finally, the ultimate layer is purpose—the engine driving every other element. The top of this trapezium is visually open, which symbolizes the idea that living in purpose is infinite. Each layer is reinforced by various habits as you move your way up from one level to the next, developing into a resilient, strong Corporate Athlete.

The Corporate Athlete High-Performance Trapezium is a multiuse tool for reflection. It is my hope that you utilize this to introspectively check in and assess your feelings, your progress, and your goals, much like a professional athlete does on a regular basis. This trapezium allows the Corporate Athlete to calibrate their approach and orientation as opposed to blindly pushing forward with little regard of the subtle changes in the environment. I truly believe that an entrepreneur's work experience goes beyond time spent at a company. This time spent can be made richer by the formation of healthy habits that benefit both the individual and the company. This graphic can be an effective tool, allowing you to convert time to enrich your work experience and giving room for reflection so you can make small, but vital, changes if and when you need to.

The concept of the trapezium takes a holistic approach to targeting the proverbial balance in your life as you work toward your professional goals. Many companies only focus on the cognitive and work bandwidth of their employees—they are not interested in how you achieve a steady approach to high performance and can, at times, even be critical of the measures you take to maintain that balance. For example, when I worked as CEO at OncoSec, I liked to run in the middle of the day. I used it as time to think and work through any of the day's seemingly unsolvable problems. I always returned with more clarity, ready to take on the rest of the day. But one of my shareholders had a problem with this routine. He asked, "How do you find time to run every day when you should be focused on the company? You should use that time to work instead of run." But I *was* working. That person simply did not understand that for me, work comes in different shapes and forms. In this case, my mental capacity was being enhanced by my physical capacity.

Similar to the parts of a perfectly calibrated F1 race car, each element of the trapezium works together to help us become as high performing as possible.

Ultimately, it is up to you if you want to operate like a tractor or a winning race car. When individuals perform to the best of their ability within the trapezium, this will naturally and positively affect the entire team. Loehr and Schwartz believe "increasing capacity at all levels allows athletes and executives alike to bring their talents and skills to full ignition to sustain high performance over time," a condition known as Ideal Performance State (IPS)[12]. Much like in sports, for Corporate Athletes to achieve high performance, setting goals and creating habits around each component are required. This allows you to enter a flow where you can work at the height of your potential for a sustained period of time.

Each component of the trapezium must be aligned in order for the IPS to occur. Many decision-makers and leaders face rising demands and feel constant pressure to get more done, causing them to split their attention and sacrifice focus, reflection, health, nutrition, creativity, and the ability to see the big picture. When we juggle multiple activities in an attempt to keep up, we become partially engaged in many things, but rarely fully engaged in anything. I could relate to this throughout different times in my career and maybe you too can also point to specific examples in your career. By reflecting on these examples, you may agree the result was perpetual frustration and a sense of hopelessness that ultimately led to burnout. What positive habits can you implement to maximize your time and energy? How can you set up your unique IPS and high-performance plan?

There are distinct differences between corporate and professional athletes. The average professional athlete spends most of their time practicing and only a small percentage of their time competing. The average executive devotes almost no time to training or revising plays and must perform on

12 Ibid., 79, 120-128, 176.

demand for ten to fourteen hours a day or more. The professional athlete has an off-season and a chance to recuperate. The average executive gets three to four weeks of vacation a year if they are lucky, but unsurprisingly, they often work on vacation too. A professional athlete's career lasts maybe twenty years, maximum. The average executive can spend four to five decades in their profession. Yet no matter how long, how often or how hard they work, both athletes and executives must consistently increase capacity at all levels of the pyramid in order to sustain high performance over time.

Ensuring all these elements are aligned requires repetition and focus. The best way to guarantee you devote time and attention to each throughout your day is by creating routines around your work and your well-being.

SUSTAINING YOUR FOUNDATION

When you are facing demands and pressures, the last thing you want to do is slow down and take a break. You might even be worried that if you do, you will lose your momentum and break your stride. More importantly, there appears to be a societal belief that taking time out for recovery and renewal is a sign of laziness or a lack of ambition. However, research suggests differently. Human beings are not designed for long periods of intense activity. Our ability to maintain focus and concentration for an extended time is limited. Although we do not readily recognize the signs of fatigue, they impact our ability to think clearly, maintain a calm focus, and control our emotions. The more exhausted we become, the more vulnerable we are to distractions and the less resilient we become. We live in an era when we are bombarded with a never-ending stream of requests, deadlines, and imperatives, all of which we deem to be important. Our brains are frequently preoccupied with work or evaluating future outcomes instead of focusing on the present. There is little downtime unless you create opportunities for it. The brain needs respite to remain productive and to generate new ideas. The body requires recovery after a significant training load; therefore, idleness is not a vacation or an indulgence—it is indispensable for the brain and mental health. The structure that provides that renewal is supported by rest, recovery, perspective, and focus, with these individual components working collectively in order to provide the necessary conditions for you to step back and view your purpose as a whole. This creates space, allowing you to make the unexpected connections for moments of inspiration to occur.

THE RENEWAL TRIFECTA

In the following paragraphs, you will find no new earth-shattering information about personal renewal. However, I would not be doing you any justice if I failed to bring these concepts to your attention. Rest, recovery, perspective, and focus are integral for balance and resilience. When the going gets tough, these are the tools that will help you centre yourself and actively recover from adversity.

What my mother told me was true: going to bed early, getting restful sleep, and waking up early is critical to my productivity. Having a ritual around rest that I attempt to adhere to daily brings a level of consistency to my routine. Although I aim to go to bed and wake up at the same time each day, sometimes life has other plans. I do my best, however, because having a regular sleep cycle impacts all other areas of life.

When it comes to workplace efficiency, nobody can work twelve, fourteen or sixteen hours straight and still feel productive. It is important to incorporate recovery into your routine in order to deal with your body's natural capacity for stress before things start breaking down. Athletes know their bodies can only work so hard for so long before they absolutely require rest. You must show your own body similar respect. If you are working long hours consistently, sporadic breaks and predictable time off for recovery renews energy, improves self-control, and helps to fuel performance. Time off can also help to stimulate creativity and heighten your awareness. Research on naps, meditation, nature walks, and the habits of exceptional artists and athletes reveals how mental breaks increase productivity, replenish attention, solidify memories, and encourage creativity[13].

Beyond the value of recovery, athletes also understand the power of a positive outlook. You would be hard-pressed to find an athlete dwelling on their frustration, anger, fear or stress before a competition. Rather, they are calm, focused, optimistic, and confident about whatever challenge awaits them. Positive emotions are the ignition for the energy that drives optimal performance. The alternative—focusing on negative feelings—only results in self-sabotage. Sustainable success requires an emotional stamina centred on positivity.

Finally, mental stamina is critical for everyone, especially those who find themselves in roles that require them to field a wide range of questions and solve many different problems every day. Complex decision-making requires attention and focus, and meditation can help strengthen both. If you are new to meditation, I suggest starting by simply devoting a designated amount of time to be quiet with yourself each day. I try to spend fifteen minutes in the morning and another fifteen in the evening on

---

13 Ferris Jabr, "Why Your Brain Needs More Downtime: Research on Naps, Meditation, Nature Walks, and Habits of Exceptional Artists and Athletes Reveals How Mental Breaks Increase Productivity, Replenish Attention, Solidify Memories and Encourage Creativity," *Scientific American*, October, 2013, https://www.scientificamerican.com/article/mental-downtime/.

meditation and doing deep-breathing exercises. That half hour each day has helped me focus my energy.

It is up to you how you make space for renewal in your life. The key, however, is to *make* space for it. There are no new and exciting ways to rest. There are apps that can help you settle into a regular renewal routine, but the one sure-fire way to succeed at this is to put away the excuses and power down. Your commitment does not have to be intense, only intentional. Even using meditation for a few minutes to break up your day can give you the recovery you need. Find your own balance and determine what works best for you.

MY ROUTINE

When I look at my daily routine, it is not that different from your typical high-achieving executive with a family in corporate America. What I am more intrigued about are the nuances that set these daily routines apart. My intention is to ensure that we are optimizing ourselves around our individual CAHPT and actively addressing any areas we need to pay more attention to. Without a doubt, our personal accountability begins with what we do every single day. Incorporating the world-class Corporate Athlete principles are key to creating a change in mindset, encouraging you to edit and adapt your routine to achieve personal success. The definition of success is a subjective one and it should not be a measure of your self-worth but a measure of how well you are dialed in to your purpose. With the help of the CAHPT framework, you will be better equipped to establish the daily supporting structure you need to meet your personal expectations of yourself.

Even though I have been waking up at five in the morning for more than half my life, it has taken me over fifteen years to refine my schedule into what it is today. Even so, I am sure that it will change in the future.

Incorporating a routine for each day helps keep me high performing and best prepared to pursue my goals. I remain systematic, with several measurements for my professional life (work), solitary time (self), and personal time that revolves around family and friends (life). We will delve into this concept of work-self-life in more detail in Principle 9. I try to make space for all areas of my life and give them attention. Otherwise, it is easy to favour one over the other.

My daily routine is heavily focused on visualization techniques, where I spend time picturing my desired future and sensing how I would feel. I set long-term goals and make my punch list to accomplish them, focusing on what the outcome would look like, success or failure. I rely on short-term goals (ninety days, thirty days, seven days, and daily) to help me achieve my objectives. My routine looks like everyone else's routine, but where I differ is my approach—I am deliberate with every single minute of the day, from the moment I wake up to the moment my head hits the pillow. If my discoveries are work related, I share this with the appropriate team member. If they are self related, I share my thoughts with different accountability partners (for example, friends) and if they are life related, I share them with my wife over coffee every morning. In my experience, I am more likely to accomplish a task if I have committed to do so by sharing my plans with someone else.

I believe success in achieving IPS hinges on the first hour after getting out of bed, otherwise, I am fighting an uphill battle all day. When we have tough mornings, it is not because of insufficient willpower. They are tough because no one teaches us how to make them easy, but there are so many tools to help simplify and bring joy to the start of your day. Here are some rules I follow that may also help you achieve a good momentum regularly:

1. Aim to be asleep each night by 10 p.m.
2. Wake up at up 5 a.m., preferably using a circadian rhythm clock.
3. Make the bed.
4. Have a morning water cocktail before coffee. I like to add a quarter of a lemon, a pinch of cayenne pepper, and salt to my water to replace lost electrolytes. I pay careful attention to food intake, but I eat what feels good.
5. Engage in movement. Sometimes, this might mean squeezing in some lunges, push-ups or burpees between meetings or simply practicing mindful breathing.
6. Be kind to yourself. Be consistent with self-care by practicing good hygiene, grooming, and dressing for confidence.
7. Check in with your list.
8. Connect with loved ones. Take the time to acknowledge them and tap into their lives. Strive to keep home and work operations separate.
9. If nothing else, carve out fifteen minutes of solitary time each day, preferably in the morning.
10. Reward yourself with a treat. (I love ice cream sundaes!)

Here is a breakdown of a typical morning routine:

- 5:00 a.m. — Wake up, followed by fifteen minutes of quiet
- 5:30 to 6:30 a.m. — Morning flow routine and workout
- 6:30 to 7:00 a.m. — Make my punch list, catch up on headlines and important emails
- 7:00 a.m. — Breakfast with the kids
- 8:00 a.m. — Take the kids to school
- 8:30 a.m. to 8:45 a.m. — Coffee with my wife
- 9:00 a.m. — Arrive at the office

My workdays vary in length but regardless of how long or short they are, I make it a point to take a break from work once I am home. In the evenings, I try to exercise again because I have to break a meaningful sweat every day, otherwise, I am not satisfied. After years of putting up with the interruptions, these days, we do not allow phones at the dinner table. My wife and I try to avoid talking shop in the house, but we make time for it while ensuring we are keeping to our pledge to the "Dhillon Family High-Performance Trapezium." Again, just as they do for everyone else, things come up that are impossible to avoid, but we do our best to focus on our time together as a family when I am at home. Depending on what is on the docket to be completed for work, I finish between 8 p.m. and 10 p.m., capping off the day with some reading or strategic planning. Fifteen minutes before bed, we usually meditate.

On the weekends, I bike on Saturday at 7 a.m. and run on Sundays by 6 a.m. The rest of the time is devoted to kids' activities, though I would be lying if I said I swear off work entirely on the weekends. There is always a bit of sneaking it in, especially on Sunday mornings and evenings, because I like to prepare for the coming week by building my checklist of tasks I aim to complete.

All of this helps me establish a ritual and positive habits, and I know if I do it week after week, it creates consistency for myself and everyone around me. This is not to say it always comes easy—a lot of work goes into it. But having a ritual gives me the opportunity to pause, look inside myself, and adapt when necessary in order to be able to achieve all my goals.

MOTIVATION-RESILIENCE CONVERGENCE

When you develop a passionate, ambitious outlook on life, it directly correlates to personal and professional success. I stumbled upon this insight through athletics and by observing other athletes over many years.

I started running and cycling seriously over a decade ago and have completed a fair amount of distance in training and races. When I first began running and cycling, I noticed that many executives and successful people from the biotech sector were also participating in endurance sports, and many approached these activities with a degree of passion and commitment that was commendable. As I observed this trend, I wondered why so many successful people were gravitating to such sports. My friend Ted Kennedy, former VP of sales and marketing of Ironman America, used to run an organization called CEO Challenges. He would host these events where CEOs could carry out their dream of competing or training with a professional athlete for a day or purchase a slot to compete in some of the most prestigious races in the world, such as the Ironman World Championship in Kona. I started to see the patterns—all these sports, be it running, cycling, golf or triathlon, were mostly individual sports; these are not what you would characterize as team sports. The other interesting thing to note is that you do not necessarily play these sports to win. They are mostly about competing with yourself, continuously trying to improve, competing against your previous bests, and trying to progress incrementally. These are not merely endurance sports but through the lens of the Corporate Athlete, they are "personal-drive" sports. Personal-drive sports are very different from recreational sports. You do not run a marathon for recreation or pleasure. Although many athletes claim to be recreational athletes by nature, I personally think this is a humble statement. There are many times throughout a marathon where you are tested mentally and physically. It is a personal-drive sport because you do it to undertake the challenge and push yourself to your limits. There is an intersection of motivation and resilience that exists here and the combination of these two driving aspects is what defines both the athlete and Corporate Athlete.

Most of us are familiar with Maslow's hierarchy of needs, a classic pyramid depicting motivational theory:

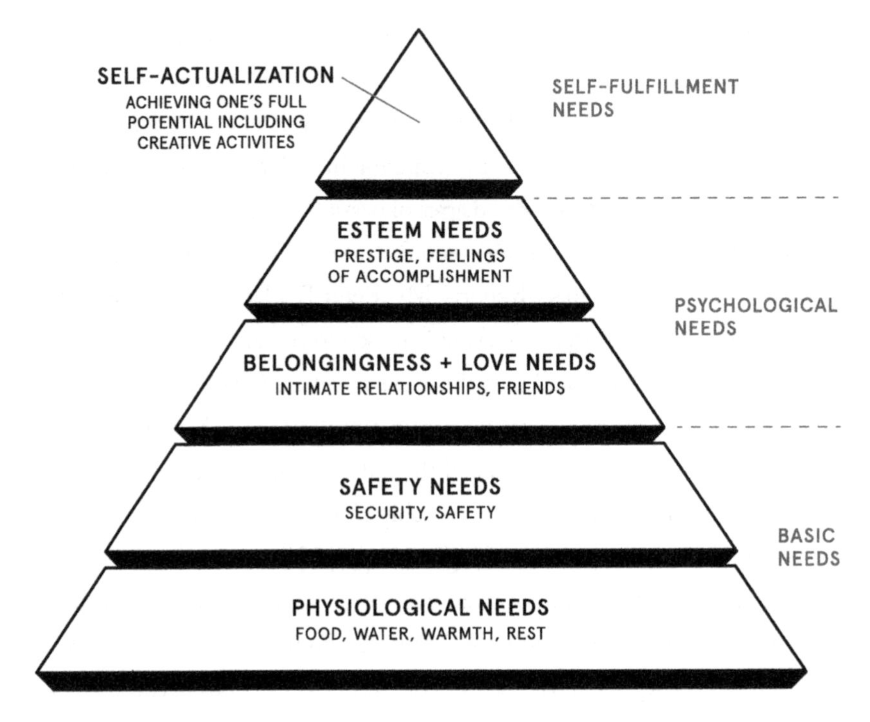

Maslow's Hierarchy of Needs[14]

Maslow's hierarchy attempts to explain our motivations in life. It starts with our basic physiological and safety needs at the bottom of the pyramid and expands vertically to self-actualization, or the achievement of our potential. These are the motivations that drive us as we attempt to ascend the levels of this pyramid. While the concept of motivation is easier to understand given Maslow's hierarchy, I wanted to understand the mechanisms of motivation in action, how motivation and daily habits are interrelated, and how they translate to personal and professional achievement. In trying to understand this, the CAHPT was born.

14 A. H. Maslow, "A Theory of Human Motivation," *Psychological Review*, 1943, vol. 50, no. 4, pp. 370–396, doi:10.1037/h0054346.

The CAHPT explains the layers that instill intrinsic motivation toward purpose, what motivates us in our careers and provides the basic requirements that serve as the engine to foster resilience. We are highly motivated by a sense of purpose at work and in life that goes beyond our own achievements. The combination of these layers represents the motivations that drive us day to day, despite setbacks. They let us focus on the "why," which extends beyond our personal needs, allowing us to remain connected to our overarching cause, vision, and purpose. Each layer works to strengthen our resolve and recovery. Therefore, motivation and resilience go hand in hand. A resilient mindset starts with a resilient regimen and the bulk of my success is based on avoiding complacency. How do I create momentum for the day? Discipline. It simply propels me through a purposeful day because it gives me a sense of duty.

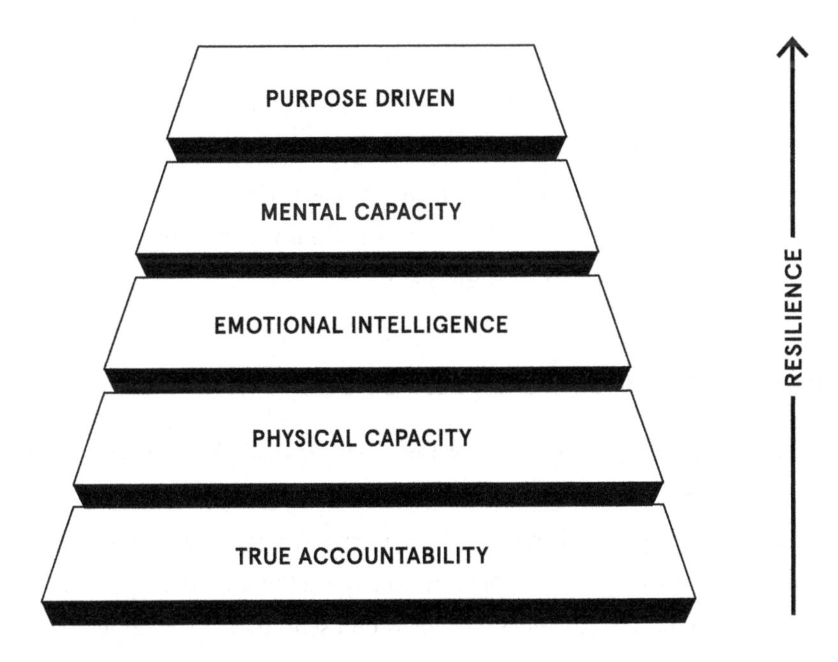

CAHPT and Resilience

There is one fundamental difference between Maslow's hierarchy and the CAHPT and understanding this difference is crucial to understanding the impact of your daily routine and your work-self-life balance: motivation. In Maslow's hierarchy, most of us have found it relatively easy to meet the physiological and safety needs at the base of the pyramid. Most readers of this book likely come from families that have provided for these needs in the early stages of their lives, and later, they have found jobs, which means they are not fighting to meet these basic needs. This means that a greater part of our lives is focused on the middle and the top layers of the pyramid. This is where the CAHPT differs. The base of the trapezium is true accountability, a need which is linked to your values, and without it, any achievement will not feel complete. This is what I like to call that internal "striving voice" because it keeps you up at night when you have not resolved something. Unlike Maslow's hierarchy, where you can satisfy the physiological and safety needs and then simply move on to the next level, in the CAHPT, the measurement of success needs this element of motivation, collectively working toward the IPS and accountability. You cannot turn off the internal striving voice. The more you align with the CAHPT, the louder it becomes and the more you are tuned into your IPS, like a top that spins forever. It becomes impossible to move away from your purpose, a sign that you are most definitely on the right path.

REINFORCING YOUR STRUCTURE

When maintaining an optimal level of your personal IPS, it helps to surround yourself with others who can strengthen and support your foundation. Alongside physical training, I try to spend forty-five minutes every week working with Justin Noppé, an experienced brain coach specializing in maximizing brain efficiency. We focus on useful tools I can apply to my work every day, including memory techniques and other exercises. This is part of the ritual I have created around the mental element of my high-performance trapezium and other members of my family have benefited as well. Not long ago, my wife and I sat down with our daughters and asked them about some areas they would like to improve academically—for one, it was reading, and for the other, it was math—and now, they spend time working with Justin every week as well. Our intention is not to upend what they are learning in school. The idea is to figure out ways we can help them strengthen their own foundations and interests by instilling in them a desire to expand their brains. It is inspiring to see children grow and witness firsthand their desire for learning new skills. It also gives them somebody other than their parents to run ideas by and to be accountable to.

In the entrepreneurial world, you are going to be constantly challenged with things outside of your expertise and you are going to be forced to think beyond your personal experiences. You will have to continue to push those boundaries and work on your weaknesses, so you are able to perform when you need to. Just as athletes would work to perfect the skills they need in order to remain competitive, you too will have to be prepared to acknowledge areas where you need improvement.

Like many athletes, you will also need a team to hold you accountable both in the workplace and, perhaps more importantly, outside of the office. My family knows the type of balance I need in my life in order to operate at a high level and if I ever forget it, they are quick to remind me. The routines and rituals we follow at home are directly linked to the strength of my foundation. Disengaging from work to sit down at the dinner table, help my kids with homework or just have a conversation with them helps remind me why I do the work I do. Continuing to enforce that work-self-life balance provides me with a deeper sense of purpose and gives me the energy I need to be present and productive for both my work and family. The more time and effort I invest into the foundation of our household, the more confidence I have to take on everything I am doing outside the home.

PASSION

Personal drive in athletics and the pursuit of entrepreneurial adventures are one and the same and by nature, polarizing—if you want to be good at either and preferably both, there is no middle ground. I have seen relationships grow stronger as well as be torn apart over someone's passion for something. Endurance athletes and Corporate Athletes tend to be deeply devoted and fanatical. These are simply the qualities needed to excel in their endeavours. Is this a strike against them? Not necessarily. These same personality traits can also lead to great devotion and dedication to a relationship and to a family. The dark side is that such individuals can easily appear obsessive and self-absorbed and many may actually be both.

When positively expressed, passion leads to the achievement of something worthwhile in an entrepreneur's career, be it a position of importance, increased responsibility or seniority. At senior levels, one of the critical drivers of success is leadership because how you engage, motivate, and lead others is a direct reflection of what you are capable of achieving. In reference to the CAHPT, the emotional intelligence and mental capacity layers bring the coach out in a leader, allowing others to learn and build their skills. Leaders who are solely focused on meeting their own objectives and who are not operating from the upper bounds of the trapezium will be ineffective and dysfunctional. Such leaders are focused on furthering their own careers and are often poor at managing their teams because they do not leverage the biggest driver for their success: their team. This can create a vicious cycle: the leader focuses on achieving more for themself, which demotivates the team, making success even more difficult for the leader and so on. I have seen these cycles develop to a point where the leader and the team become competitors. Often, these cycles only end with the exit of the leader or, sadly, the exit of the entire team.

At a Corporate Athlete level, operating from purpose and not purely focusing on your own achievements is easy to say but difficult to do. If you are reading this book, you are probably a high achiever, and this has led you to experience both career and personal growth. How can you fulfill your own goals without being blinded by your own success? A world-class Corporate Athlete manages to reconcile and overcome this contradiction, going on to achieve greater successes by bringing others along with them.

When directed inward, passion can be ill-expressed in a corporate setting. It is most functional when it is used to facilitate leadership as an impetus to achieve shared goals as opposed to fueling personal objectives. When leaders are fortunate enough to gain capital, they should not hoard it but reinvest it back in the company in order to elevate everyone who has contributed to the success. This is true leadership when the leader invests in the people who have enabled said success. The allocation of power amongst the team does not dilute it; in fact, it only increases it because the collective has that much more strength. For example, Ikea famously finds creative ways to express gratitude to their employees, devising various strategies to show appreciation to those working hard, day in and day out. In 1999, the company divided a day's worth of generated profits among its 40,000 employees as a means of rewarding and retaining workers. It was unheard of at the time, but it showed genuine innovation by nurturing those from the ground up. Success does not come from just one person and a true leader will recognize this and act accordingly.

We are in this together, doing different things in order to support each other. We understand that accountability and purpose is much deeper than just looking out for ourselves. Looking for opportunities to elevate others is perhaps even more important than seeking ways to help yourself, and the act of looking outward nearly always comes back to benefit you in ways you could never anticipate.

□

WORLD-CLASS CODA

- When you are learning any new skill, repetition is critical. The resilience regimen is a long-term plan, and a capacity for resilience is critical to maintaining high performance.

- Establishing a routine centred on a form of structure or discipline establishes a positive energy cycle, and creating habits driven by deeply held values will help you to effectively manage energy. When you clearly define what is most important to you and align your daily actions and behaviours with your highest values, your calendar will reflect your priorities.

- Explore ways in which adequate recovery and renewal can be included into your routine.

CATAPULT FORWARD

- If you were to rearrange your work environment to cultivate greater resilience, what would you change?

- Do you have adequate time off to stimulate creativity and heighten your awareness to self-check your accountability toward purpose?

- How is stress affecting you? How is this either building or eroding your resilience? What are you willing to do about it?

There is power in numbers: an extra set of hands; another set of shoulders to climb on. What you build alone can be wonderful, but what you can build with others can be extraordinary. When you are in good company, sharing a collective vision, your reach is profound. When you look beyond yourself, roll up your shirt sleeves, and lend a helping hand, this is the kind of service that is radical and transformative. This is what uplifting humanity looks like. This is true philanthropy.

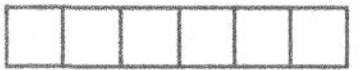

PRINCIPLE 5

DRAFTING

"Alone, we can do so little; together we can do so much."
—HELEN KELLER

I have a dream race on my bucket list called the Comrades Marathon. It is an ultramarathon, approximately fifty-five miles, which is run annually in the KwaZulu-Natal province of South Africa between the cities of Durban and Pietermaritzburg. It is known as the world's largest and oldest ultramarathon race. The direction of the race alternates each year between the "up" run, starting from Durban, and the "down" run, starting from Pietermaritzburg. The first race, which was the idea of World War I veteran

Vic Clapham to commemorate the South African soldiers killed during the war, was held in 1921. Clapham, who had endured a 2,700-kilometre march through sweltering German East Africa, wanted the memorial to be a unique test of the physical endurance of the entrants. The constitution of the race states that one of its primary aims is to "celebrate mankind's spirit over adversity," and it is said that no runner is left behind, symbolizing the camaraderie required to complete this feat of endurance.

With the dream of Comrades etched in my brain, I have no doubt subconsciously and consciously created a sense of camaraderie in everything I do. The support and positive energy of friends, work, family, and coaches make anything possible because this is the scaffolding for discipline. In cycling, a phenomenon known as drafting occurs when a cyclist moves behind another rider and into an area of lower pressure. Essentially, the cyclist leading the pack takes on the headwind in order to assist his teammates to reduce energy cost. This is not a one-way relationship either; the cyclist at the front also benefits from drafting, as the other riders eliminate turbulence from the airstream moving over them. This is a perfect illustration of pure synergy: both parties are helping each other for the greater good.

MAKE THINGS EASIER

A big part of success for a Corporate Athlete is to make things easier for people. This is not meant to be confused with making things *look* easy, when we all know how much time is invested in perfecting and excelling at a skill. Making things easier for others and, ultimately, ourselves is key to retaining business, clients, and relationships. If there are three valuable takeaways I have learned as CEO to facilitate efficiency and loyalty, it is:

1. Remove obstacles for your team
2. Keep the company and team moving forward, and
3. Ensure the company remains capitalized to achieve points one and two.

To give you some forethought, I will give you a few end-of-the-chapter Catapult Forward questions in advance:

- Do we make it easy for our teammates to do their best at their jobs?
- Do we make it easy for customers and other stakeholders to do business with us?
- Do we make it easy for customers and other stakeholders to give us useful feedback?
- Do we make our product or service easy to access, understand, use or buy?
- Do we make our product or service easy to recommend?

Whether you are in a sports team or in an organization, variations of these questions can apply, and each time we have a problem, we must ask ourselves, "Can we make things easier and, if so, how?" This question applies to every facet of a business.

As a leader of a business or a captain on a sports team, you are ultimately responsible for your organization's ability to acquire and keep customers or winning for your loyal fan base, which many would say is the fundamental reason why any business exists. Ask yourself this simple question: *Would it be a powerful competitive advantage if we were considered the easiest to do business with within our market?* Definitely! Being the easiest to do business with is optimal in terms of being a decisive differentiator for today's customers when it comes to absolutely anything—banking, professional services, booking an event, buying groceries or even visiting your physician for a check-up via telemedicine. When you reduce the effort required by a customer, you automatically add value. Consider companies often cited as disruptors in their industry—what makes them so popular?

- Tesla made it easier and more affordable to own an electric vehicle.
- Uber/Lyft made it easier to catch a cashless ride.
- Amazon made it easier to buy virtually anything and receive it on your doorstep within hours.
- Apple made it easier to navigate the intricacies of technology.

As a Corporate Athlete, if you are intentionally leading your organization in an effort to make things easier to understand, access, use or buy, then you are succeeding at one of your most important jobs. I always try and apply the standard of making things easier to everything I do in my business.

MOST VALUABLE PLAYER

Over the years, I have observed my own evolution as a leader. In the early part of my career, I was an achievement junkie. As I started to ascend to middle management and lower leadership positions, I started seeing the earliest signs of dysfunction in my leadership style. The self-prognosis was that despite a lot of achievement, doing our jobs, and getting what needed to get done, my team was not necessarily happy or motivated. In fact, I remember a particular incident while visiting one of our divisions in Oslo, Norway, where I met with the CFO and I vented frustrations with him about the team. Something was clearly not working and yet I did not have the foresight or the solutions to fix what was wrong. As a result of not knowing what to do, I continued to push the team with no real agenda or understanding of what was missing.

BREAKAWAY

ALL FOR ONE, ONE FOR ALL

If you are a runner, cyclist or swimmer, you spend many hours alone training in remote, unpopulated areas. Endurance training can be a solitary undertaking, yet it also can introduce you to people and worlds beyond your imagination. Some of the greatest and deepest joys in my triathlon career came from moments I have shared with people I met along the way.

You do not have to be an endurance athlete to take advantage of the social rewards of running or cycling. Try running with a friend on some of your routes or get involved with a weekly running, cycling or triathlon club. Do something for a sport you are passionate about that does not involve the sport—work at the finish line or at an aid station, coach others or join others on their journey through a half-marathon or marathon. These are all things I have done, and every single one has created great ways for me to participate and give back to the sport that has given me so much.

There are many ways to prepare to run a marathon and many training plans do not necessarily require you to have an accountability partner, rather they ask you to keep your head down and nose to the grindstone. But there are also many successful ways for people to train alongside others. I encourage you to surround yourself with supportive people, whether it is your coach, your teammates, your friends, your family or a new running partner. You, and the people around you, will benefit so much more from the camaraderie.

In a twist of fate, my achievement orientation and competitiveness spurred my interest to run a half-marathon and I enrolled. To bolster team morale, I suggested we make it a company-wide challenge to sign up for either the half-marathon or the 5K run, which was happening at the same time. As the leader of the initiative, I wanted to encourage people to join by leading from the front. I hosted lunchtime and after-work runs, provided pseudo-coaching tips, and even obtained approval from management to install a gym at the office. After this year, the 5K race became an annual event for the company and, personally, I was very happy with my own improvement in running and dedication to the sport. But more than my personal success in running, what I noticed was a gradual shift in my leadership style. My desperation to succeed at work had reduced. My own achievement started to become secondary. The less I focused on my own wins and celebrated those of the team, the more success I achieved at work. I secured some of my greatest career achievements during this phase because my team was completely engaged. We worked together with passion and for a shared cause. Mentors like my former CEO Joseph Kim, who had known me for a long time, could clearly see the change. I had transformed into a better version of a leader in the phase of four years.

During that period, I led a merger of two companies that catalyzed one of the strongest DNA-based vaccine companies in the world, Inovio Pharmaceuticals, which, at the time of writing this book, is conscientiously developing a vaccine for COVID-19, joining other vaccines and therapeutic contenders that will collectively resolve the virus in the future. The leadership trust built with the Inovio team also led to the creation of technology that allowed me start OncoSec, a cancer immunotherapy company. During this same period, I also ran three marathons, nine half-marathons, completed two half Ironman competitions, and an Ironman. I have no doubt that this change in my leadership effectiveness at work was accelerated by my passions outside of work.

CHEERING SQUAD

In my first years of high school, I was already quietly establishing a reputation for being a star student. By the time I was in ninth grade, I had gained a certain standing for being committed to academics with a full after-school schedule of extracurricular activities. My peers knew it, the teachers knew it, and the year prior, my guidance counsellor presented me with the Grade Eight Outstanding Student Award. I loved having all my hard work and drive recognized and by the next year, I wanted that award again. My plan was to just keep doing what I was doing, as it was clearly working.

I knew my eighth-grade counsellor would sing my praises to the new counsellor I was assigned to in ninth grade. I also knew the new counsellor was acquainted with my family because she had been both my mom's and my uncle's counsellor in prior years. I might as well have collected the award on day one of ninth grade!

At the awards ceremony at the end of the school year, I was sitting in the audience with my smile barely contained, completely confident that I would be a repeat winner, when the Grade Nine Outstanding Student category was announced. I was almost out of my seat when I heard my counsellor say, "And this year's winner is: Nelson Siu!"

I was stunned. I had been on a roll the whole year—I even had an "in" with the counsellor! Had I not worked hard enough? What did Nelson do that I failed to do? How was I going to explain this to my parents? I was indignant, shocked, and embarrassed, in complete disbelief. After wrestling with this injustice internally for a few days, I decided I needed to go to the counsellor's office for an explanation and for my own peace of mind.

"Look," I said, "I need to know why Nelson won the award and not me."

Instead of dismissing me or calling me out on my immaturity, she sat there quietly and listened. She let me vent and air my grievance. And when I was done, she had a question of her own.

"Punit," she said, "what are your goals?"

"Well, I have some big footsteps to follow," I said. "My dad has a master's in engineering. My uncle is a former champion wrestler and now he is a doctor. He is the star of the family. Honestly, I feel like I have a lot of pressure to be just as successful. I am the next person in the entire family who is slated for higher education. There are a lot of expectations on me."

A small frown formed across her face.

"All of that is just about putting pressure on yourself," she said. "You have been a wonderful leader and a big supporter and encourager of your fellow classmates and those are the things that will serve you really well, not some shiny object that is going to sit somewhere and collect dust. In the end, your ability to elevate those around you will be far more valuable."

I was speechless. I had never thought about the possible impact of my behaviour that way. I had never really thought about anything, except for how all my hard work would lead to eventual glory. Yet in that moment, I began to consider the importance of opening doors for other people. I saw the potential for how a positive outlook could spread. If Nelson was recognized today, think of how that would change him. If he used some of that energy to elevate those around him, the impact could reverberate through the whole school.

It was a big conversation, particularly for a somewhat immature adolescent, but her point is one I have never stopped living. That pivotal discussion was the impetus for my entrepreneurial spirit, which is rooted in my interest in the growth of others. Like most entrepreneurs, I thrive when I am solving problems. This is why Nina and I remain committed to devoting time to volunteering and community involvement, whether it be through participation on the PTA, YELL or any of the other organizations we support. Look at all the successful entrepreneurs in the world—most donate generous portions of their income to organizations dedicated to doing something better for the world. When helping others is at the core of your purpose, it is impossible to spend billions of dollars on things that just do not matter. The Bill and Melinda Gates Foundation is a perfect example. Bill Gates is now 100 percent committed to addressing global issues, and he is putting not just his energy into it, he is putting capital behind it. He understands that the greatest form of wealth comes from helping other people solve problems.

Investing your time and energy in others helps you become more confident in your own purpose. It reaffirms your vision by giving you fresh energy for pursuing your passion and your goals. It makes you continually work harder and perpetuates your drive to do anything necessary to get to your desired outcome because you would not be holding yourself accountable otherwise. Furthermore, when other people are depending on you, you are less inclined to give up. When that double accountability occurs, the benefits can far exceed expectations.

Nelson Siu went on to work for Bill Gates at Microsoft, playing a big role in designing the popular email program we all know, love, and use, called Outlook. His work is helping millions. The teachers saw potential in him that day, something I was too young and inexperienced to recognize. But they did and they were absolutely spot-on—he *was* outstanding. We all have essential roles we play and just because we fail to get what we want does not mean we fail to get what we need. I was meant to be his cheerleader that day and he was meant to be on the stage. We all have our turn and if we can spend the time supporting and elevating those around us when the spotlight is not on us, we can still derive important life lessons in those experiences. In hindsight, I know I absolutely did.

WIN-WIN RELATIONSHIPS

Throughout my career, I have found that when others are invested in the same purpose as I am, I become invested in their purpose as well. Accountability between both parties always leads to something even better than we ever could have done on our own. In these relationships, I need that person's help to reach my goal just as they need me to reach their objectives. Individually, we can complete the task alone, but we are going to get it done faster, more efficiently, and with a higher degree of success by doing it together. Here, the idea of community is all-encompassing—it can support an individual's ability to achieve as well as rally together to carry out an ambitious goal.

Successful entrepreneurs know one person can only go so far. Working with a team with a wide range of knowledge and expertise creates an environment where people are nurturing one another, resulting in strong social capital. The exponential effect of such behaviour helps to frame the policies and practices of a company and when the entrepreneurial spirit is magnified, change inevitably happens.

BETTER TOGETHER

At my former company, OncoSec, the outcomes of such a culture led to significant advances in cancer immunotherapy. When I launched the company in 2011, we started out treating tumours on the surface of the skin in stage three and stage four melanoma patients, but we ended up fully collaborating with one of the biggest pharmaceutical companies in the world, Merck. Our drug, Tavo, worked in combination with their drug, Keytruda, to treat Stage 3 and Stage 4 melanoma patients with no other treatment options.

In the company's early days, I bet on a technology that had a clinical study in the *Journal of Clinical Oncology*, a peer-reviewed publication, which showed a 53 percent overall response rate in melanoma patients of a drug that had not been used in an intratumoral setting. Although very compelling and published in a reputable peer-reviewed journal, the study received some criticism because the findings were not based on a comparison study but rather an investigator-sponsored study, meaning it did not offer a clear statistically significant signal of the therapy without any comparison against a placebo, as it was based on a small sample size. But seeing great potential in this work, I asked Dr. Daud, the lead clinician, how he felt about furthering it. My plan was to take it through the proper drug development process, seeing as he did not really know where to start as far as that was concerned. While he was among the top five melanoma doctors in the world and clearly knew the science of the drug, he did not consider things from a business perspective.

In fact, Dr. Daud could have cared less about the drug making him any money at all. He was far more concerned about the drug helping patients because that was his true passion. So, we had two individuals: a doctor who was saying, "My focus is not on business, but I care about the success of this drug because I believe it works," and me, who was saying, "Right, but this also makes very good business sense." Fortunately, we ultimately had the same goal of getting the drug to patients who needed it and while he worked on perfecting it, I wrapped it into something digestible for investors, and the company was formed.

Dr. Daud could have taken an equity position in the company and become incentivized not only by the drug going through the development process and eventually getting commercialized, but also by profiting financially along the way. But he was not interested in the money. He is a perfect example of somebody who cares most about helping others. My partnership with him is the ultimate example of a win-win relationship—I would not have been able to start my company without him and he may not have ever seen his drug developed without me. We were both striving toward the same objective and, in the end, it opened new doors for both of us. Developing that technology for late-stage melanoma patients has been an astounding success, especially in a combination therapy with standard of care for second-line patients who have no other treatment option, where the results for these patients are showing over 30 percent response rates, which is extremely promising. At the time of this writing, the combination therapy of Tavo and Keytruda is completing the last stage of a registration clinical trial before it may be commercialized, and Dr. Daud and I remain very close friends.

THE RIPPLE EFFECT

I witnessed the possibilities of effective collaboration and cooperation among a diverse group of contributors at an early age. My high school was situated in a neighbourhood heavy with gang activity. A drive-by shooting happened at the school the year before I started there and when I was in eighth grade, an English teacher was murdered. Fights were rampant, both in the hallways and in the neighbourhood streets. There was no escaping the violence, but there was no other choice because this was our school. Even though I was, at times, scared to even go, my parents could not afford to send me to private school, so we dealt with it as best we could, just like everyone else. But at a certain point, a group of administrators decided that just dealing with it was not good enough.

Over the next five years, this amazing group of educators worked together to mobilize a community. They appealed to alumni who ended up creating an endowment to invest back into specific programs for the school from which they had graduated. The administration looked for ways both big and small to connect with every student and break down barriers. They were on friendly terms with both the gang member and the star pupil. There was a lot of energy and camaraderie in the hallways, which were lined with posters promoting positive messages. It sounds cheesy, but somehow, all these things added up to an optimism that spread throughout the school, and we saw a significant decrease in violence over time.

When you break down stigmas associated with trying to connect with people, the results can be mind-blowing. It was important to witness the principal getting to know kids from every clique, from the athletes to the model students. It made you think, "What's going on here?" Then you realized, yes, the most unlikely person in school is indeed friendly with the principal and, perhaps even more absurdly, that was the norm.

With the old barriers removed and a fresh environment in place, everybody united around an interest in serving the community. Slowly, wild things started happening. We became one of the top schools in the entire Vancouver district, gaining recognition for our dedication to community service, innovation in the engineering program, and talent in the music program. We participated in canned-food-drive competitions among other schools and each time, we won by a landslide because we had participation from every single student. We had a very approachable student council, which led by engaging leadership, transcending social circles, and bridging gaps between the grades. We had a revered drama program. We had wonderful trades and apprenticeship programs. Members of our choir were invited to sing backup for Shania Twain at one of her concerts when she performed in Vancouver. Everyone felt like they were part of something special, and our collective ecosystem continued to both evolve and flourish. Quite simply, everyone mattered. Despite the rough reputation of our school, which often drew concerns from my parents, the fact that many students were involved in extracurricular activities set a great example for me. I often looked to them to show my parents that the school was indeed safe and that these activities fostered confidence.

Last year, I spoke with my dear friend Vina Gandhi about her high school experience. She was the president of the student council when I entered eighth grade. I was not surprised to learn that she deliberately made an effort to cultivate positive energy and engagement throughout the grades, ensuring the younger students were taken under the wings of the older students, thereby establishing a precedent for students to mix and mingle with each other, regardless of age or grade. When she first joined student council, it took some convincing for her traditional and conservative parents to understand that it was a positive extracurricular activity that did not take any time away from her studies. By showing her as an example to my parents, they slowly started to understand that there were merits outside of pure academia, and I was thrilled to have a little more freedom for after-school activities.

As a fourteen-year-old at the time, I was not fully aware of the massive significance of what I was watching unfold around me, but I did recognize that cooperation between the student body and the administration had transformed my school into a different place. It was like the turnaround of a company. We succeeded because the stakeholders—the parents, the teachers and the students—were more engaged than ever. By the time I graduated, I felt like anything was possible. No perceived disadvantage could hold me back because I knew it was just a matter of looking at it differently and figuring out what to do to make it work. Clearly, that leadership knew what they were doing and the collective reinforcement from the principal and everyone down the chain impacted everything, including that invaluable advice I received from my counsellor about focusing more on my ability to help others than on some award.

The entire experience informed the kind of leader I wanted to be because it taught me how to mobilize people in order to get real, sustainable results. In the case of my school, we mobilized our mentors, the teachers, our peers, and the community. In doing so, we bridged a gap between young people and leaders so we could work together toward addressing the issues that were draining the school of its potential. Everybody was aware of the problems that needed to be solved, but we also knew there had to be a way to fix them. It took everyone guiding one another to those solutions to implement something successful. It showed me the power of working with people to get something done. You can only achieve so much by yourself and, and if you really want to see an exponential change, you need to think about what might be beneficial for the other person as well to create that reciprocal relationship.

This is a concept my peers and I wanted to make sure future generations never forgot. Years later, after my counsellor passed away, Tu, my best friend since eighth grade and a close colleague of the last ten years, and I created a memorial scholarship in her honour, which is given to students who display leadership by supporting their peers and opening doors for others. I remained in touch with my principal, Tom Grant, who became a good friend and played an instrumental role during the start-up of YELL Canada, hosting a school district-wide meeting where our team could pitch our aspirational plan. My connection to my high school remains long after I left, but the one thing that has not changed is my priority of uplifting others and serving our community. This is one thing I could never unlearn, even if I tried.

THE POWER OF COLLECTIVE UNITY

As an entrepreneur, you can always find opportunities to open doors for others. Especially early in their careers, it is easy and tempting for entrepreneurs to have a "take, take, take" mentality. But if you reverse that thinking and approach every interaction as an opportunity to elevate one another, you end up gaining much more. Because at the end, entrepreneurship is really not about you. It is about everyone and everything else. It is about your community and helping the people around you realize their potential and using that potential to affect change. This is where the human aspect of the interaction becomes critically important. It ends up being less about the supply and demand you are creating in business and commerce and more about the collective unity and what you can achieve from this space.

Similarly, effective entrepreneurship requires knowing the capability of a collective community that is trying to do what is right. This is easy to do in my field of life sciences, where most people are working on a life-changing drug for an unmet medical need. But in any industry, you need to make sure your business works and is profitable while simultaneously adding value to the world. This will take a lot of collaboration. For my company, it might mean sharing our intellectual property. It might mean spending more money on research with university collaborators to further research and seek a publication before obtaining broader investor support on an early hypothesis. It might mean sacrificing some things along the way in order to get to that ultimate outcome. No matter the scenario, we are never going to figure it out all by ourselves. Even if we do not have a single drug to sell until ten years in the future, we have to make sure we are holding ourselves accountable toward that end objective of getting the drug approved. The only way we can do that is by working together.

MEETING FUTURE LEADERS

In my life, a perfect example of the power of collective unity has been my relationship with Amit and Rattan, my fellow cofounders of YELL Canada. I met them in my early thirties via social media, when a member of my public relations team came across a community-led young entrepreneur competition, TYE Young Entrepreneurs, hosted by a group of successful entrepreneurs and executives in Silicon Valley. They had chapters around the world, including one in Vancouver, and my colleague thought it would be the perfect pairing. I reached out to Amit and Rattan on Twitter, and we hit it off instantly. Right away, they enlisted me as a mentor for one of their teams in this competition. In my very first year as mentor, my team ended up winning the world championships, which consisted of finalists from all of the different chapters of TYE from around the world. Everyone was ecstatic and I remained as a mentor for the following year, where we won the world championships for the second time. It helped that in both years, the ideas centred around medical devices, making my career experience very relevant.

Shortly after that victory, I was in Los Angeles for meetings and during some downtime, I decided to reach out to Amit and Rattan to see if they wanted to spend some time together the next time I was in Vancouver. Until then, all our interactions had been so focused on the program that we had never really had any time to socialize. In fact, in the first year of mentoring, I would often meet with my team in person in Vancouver, but rarely with Amit or Rattan, who were the cochairs of the program. That summer, a new Bond movie was coming out, so I decided to host a Bond Night, complete with tuxes, a limo, and tickets to the sold-out premiere. It was a really fun evening. After dinner, I pitched them an idea. We had been experiencing such massive success with the existing TYE program, I wanted to do something similar across all of Canada. That conversation was the beginning of the formation of YELL Canada.

By that point, in addition to being great friends, we were committed to our shared goal of helping students and bringing different communities together. Our partnership shows that you never know who you are going to meet, and how they are going to impact you. The entire experience has given us a platform to do something really meaningful from a purpose standpoint. When living your purpose, it does no good to stand on the soapbox and only talk about your vision—if you really want to implement it, you need to become operational and have a structure around it. In this sense, YELL has been an ideal outlet for all of us.

YELL is preparing students for a radically different future. It is focused on entrepreneurial thinking, where youth can perfect the art and practice of solving complex problems. The organization is committed to creating an experience for students based in courage, curiosity, compassion, and integrity. Amit serves as YELL's managing director while I, Rattan, and Dr. Sarah Lubik of Simon Fraser University serve as board members. As of 2020, we have had nearly 1,000 students graduate from the YELL program. The program is funded by a combination of government and corporate sponsors as well as private donors and is offered to 11th- and 12th-grade high school students across the Province of British Columbia. Soon, the program will expand across Canada. We have been very fortunate to leverage the experience of the collective community to obtain high school credit for the course and recently partnered with Simon Fraser University, making YELL the first high school entrepreneurship course eligible for university credit in Canada. We are now introducing programs for younger students as well. We have a very active and engaged YELL alumni group that continues to grow worldwide. YELL exemplifies community engagement by bringing in nearly 200 speakers, mentors, companies, and organizations to support youth in classrooms each year. These classrooms have become so interactive and inspiring for the students because they feel like they have a connection with somebody who is an expert in their industry. Whatever they are thinking about accomplishing is now an attainable possibility. Every year at the YELL Venture Challenge, which is an annual business plan competition and a culmination of their classroom and extracurricular experience throughout the year, they astound me with their maturity and preparedness. We ask very tough questions, and some of these teams have actual appendixes prepared so they are ready to answer anything we throw at them.

Every person has several distilling personal experiences that influence them in a meaningful way. Every person has a unique experience and story, including important people who have touched us along the way. This is the one thing we all share: we all have someone in our lives who has influenced us to become the best version of ourselves.

I encourage you to take a moment to reflect on your own path: Who was the person who said something that really stuck with you? What idea or words of wisdom did they share with you that helped shape who you are today? When we share our experiences, we leave a piece of ourselves with others, and this can have a profound impact on how they move forward in their lives.

BREAKAWAY

WISER, NOT OLDER

At every Ironman, a pre-race dinner with all the athletes is held the night before the race. Everyone has an opportunity to meet each participant, and the organizers like to highlight the stories and experiences of some of the extraordinary people in the room. Sometimes, it is the oldest person who has ever completed the competition; other times, it is someone who has competed in an Ironman every year for the last forty years.

These dinners have taught me the immeasurable value of resilience as we age. In 2014, I spoke with an older athlete in the hotel lobby who had competed in over thirty Ironman races. When I asked him why, he candidly responded that he enjoyed breaking his own personal record every year. In fact, he found the training to be harder and more gruelling each time, but despite the increasing physical and mental challenge, he found the experience to be incredibly and equally rewarding.

Other racers I have met have all told me the reason they undertake the Ironman is simply for the "love of the sport." It simply makes them feel invincible. They love that they are still able to push themselves and reach new limits. They taught me that everybody goes through the same mental gymnastics, the same pain, and the same nervousness at the start line and they all wade through their personal battles to get to the finish line.

Some of these athletes are in their late sixties, seventies, and even early eighties. In 2018, at the age of sixty, Will Turner had competed in sixty Ironman/Full Distance Triathlons in the course of one year, setting the Guinness World Record for the most Ironmans completed in a single year. In 2016, Ed Whitlock completed the Toronto Waterfront Marathon in three hours, fifty-six minutes, and thirty-four seconds to become the oldest person at 84 to run 26.2 miles in under four hours. Physiologically, they may be older, but aside from the biological age of these veteran racers, mentally, they are no different than other high achievers of any age. They are constantly pushing their own boundaries and have an even deeper sense of resilience, an urge to prove themselves repeatedly regardless of age. Their accountability is inspiring. They are never holding themselves back from the next goal and they feel exponentially stronger from each accomplishment, enough to come back and do it again. For them, there is no single finish line, but many finish lines.

Young athletes can learn a lot from the experiences of the older generation. I have met older athletes who do not have a coach, follow a special diet or wear a heart-rate monitor or any other tracking gadget. Their racing approach is simple and pragmatic. Their focus is on renewed purpose, understanding that challenging the human body should never be capped at a certain age.

People who have experienced life in this way are almost always interested in bestowing their knowledge. People naturally want to inspire others. Reach out to older athletes and see what they have learned from their personal journey. If anything, they will help you to visualize your own future.

THE NOT-SO-SECRET TO SUCCESS

Talk to any business leader, and they will tell you the most important thing about business is relationships. As social beings, we require meaningful interaction and collaboration to form our identity and facilitate personal and professional growth. At YELL, we emphasize this concept through mentorship programs that allow students to develop an idea and build it into a viable business venture while creating meaningful and lasting relationships along the way. Our creative approach to education begins with a simple shift of power from teacher to student. Through active mentorship and collaboration, we create a self-driven learning environment that provides students with the skills necessary to succeed in the modern economy and the awareness that their personal growth and success is made greater by the success of others.

We do this because we know the only way to become the best version of ourselves is to invest in and engage with others. Their perspectives, experiences, and visions of what is possible will inspire and shape you in ways you can never anticipate. The only way to live with real purpose is to recognize the value in community and contribute to it while simultaneously allowing the people around you to propel you forward in everything you do.

□

WORLD-CLASS CODA

- Most of the time, it is not just about you! While it is admirable to strive to be the best, it does not serve anyone but yourself. Widen that scope and start to think of who surrounds you, and instead of competing with them, support and root for them. By uplifting others, you start to make space. Remember, there is room for everyone on this stage.

- Collaboration is key to the advancement of a shared cause. When you hit a dead end or when a door closes, your peers may be the ones who can open the window for you. Never underestimate the efforts others can bring to your team or your ultimate goals.

- As important as your personal goals are, never lose sight of the wider world and what you can do for your community and beyond. Think outside the box and pinpoint what is missing in your community, finding solutions to these problems. Think about the generation that comes after you and the one that came before you and how supporting people can benefit the world as a whole.

CATAPULT FORWARD

- Are you able to make difficult things easier? What do you do that keeps you from getting out of your own way? How do you self-sabotage?

- Can you think of ways of uplifting your colleagues that would take competition out of the equation?

- What charitable causes are you deeply passionate about and what can you do to make lives easier for people in your community?

Sometimes you just have to step back—to take a beat, a breath, a better look. While it is admirable to strive for perfection and pay attention to the little details, you also need to know that when you have done your best, it is time to let it go. The journey is long. The periods of struggle should be balanced by periods of peace, of appreciating all that you have accomplished and preparing for the road ahead.

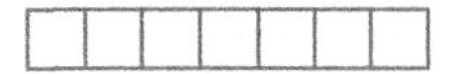

PRINCIPLE 6

VISTA

"Often it isn't the mountains ahead that wear you out, it's the little pebble in your shoe."
—MUHAMMAD ALI

During my coaching career, I witnessed firsthand the gruelling experiences of many elite swimmers and rowers who put an inordinate amount of stress on themselves, both physically and mentally. Over the years, they were socialized to accept and blindly trust that the training process was a test of not only their physical prowess but also their mental agility. Although experiencing stress is a normal byproduct of competitive sports, it is well documented that chronic stress is detrimental to athletes' well-being, inevitably leading to burnout. The amount of stress placed on overexerted athletes over an extended period of time can wear away at the psyche of athletes, myself included, an unfortunate cause and effect of constantly pushing in overdrive.

It is normal for competitive athletes to experience symptoms of fatigue, especially after a long training camp or after an extended cycle of training. After short recovery periods, the motivation for more training and new competitions quickly returns. However, a small number of athletes (1–2 percent) will experience a more serious and chronic state of exhaustion that eventually can lead to a total withdrawal from the sport, which is indicative of burnout and is a consequence of prolonged stress[15]. In sports, burnout is generally comprised of three different factors: emotional and physical exhaustion, a reduced sense of accomplishment, and sport devaluation[16,17]. The first dimension is characterized by the perceived depletion of emotional and physical resources beyond that associated with training and competition. The second dimension is characterized by a tendency to evaluate oneself negatively in terms of sport abilities and achievement. The final dimension reflects on the development of a cynical attitude toward sport participation. When viewed in light of the world-class Corporate Athlete, it is easy to see how these very factors can translate over to the business world. There may often be a discrepancy between the individual's expectations versus the reality of the situation. The satisfaction of core human

15 Henrik Gustafsson et al., "Athlete Burnout: an Integrated Model and Future Research Directions," *International Review of Sport and Exercise Psychology*, 2011, vol. 4, no. 1, pp. 3–24, doi:10.1080/1750984x.2010.541927.

16 Henrik Gustafsson et al., "Athlete Burnout: Review and Recommendations," *Current Opinion in Psychology*, 2017, vol. 16, pp. 109–113, doi:10.1016/j.copsyc.2017.05.002.

17 Thomas Raedeke and Alan L. Smith, *The Athlete Burnout Questionnaire Manual* (West Virginia: Fitness Information Technology, 2009).

needs is related to one's purpose, autonomy, and competence in their profession and is fundamental to one's well-being. When Corporate Athletes are constantly working overtime and cannot find balance in other parts of life, they can easily begin to feel frustrated by the lack of achievement and unmet core needs. From here, it is a short hop, skip, and jump to feeling pessimistic about their performance, career, and overall goals.

Athletes are particularly vulnerable to burnout syndrome because commitment is a hallmark of athletic success. When athletes are physically and mentally depleted, this commitment gradually develops into apathy, which may ultimately result in them giving up the sport altogether. It has been suggested that burnout is caused by the relentless pursuit of success. It appears that people with high ambitions and "overcommitment" are especially susceptible, but at the same time, this motivated quality is something that is extremely desirable in sport. This double-edged sword is one that both athletes and Corporate Athletes need to keep in balance, as the very characteristics that enable them to succeed may also cause them to inadvertently fail.

WIN WHEN IT MATTERS

If burnout is a state we need to mitigate before it happens, one of the key factors that will ensure we can maintain moderation in our career is pacing. At the beginning of most athletes' careers, the direct trajectory of ascent is thrilling but unsustainable. There is also inevitably a portion of their career when things plateau. How can athletes capture the energy experienced at the beginning of their career as a means of sustaining themselves through the parts that are challenging or when success feels out of reach?

Whether you are a professional athlete or a Corporate Athlete, there is a finite time for your career. A professional athlete may have a career for fifteen to twenty years, while a Corporate Athlete may roughly spend thirty to forty years in the workplace. In either case, it is important to internalize that the career is a marathon, not a sprint. A person's career can be broadly split into two halves of a marathon: the first and the second. The length of the two halves does not have to be proportionate. For the professional athlete, the first half may be directly related to their success in their sport and the second half dedicated to life after professional sports, as an entrepreneur or coach. For the Corporate Athlete, the first half is usually dedicated to striving toward significant career achievements and the second half is when you have found your stride and where true success is created. In both examples, we can conclude that a professional athlete and world-class Corporate Athlete are dedicated to their passion, taking advantage of their natural abilities and raw talents, working alongside coaches to help them maximize their potential, and establishing a robust structure with self-accountability.

In the first half, it is easy to conclude—barring no unfortunate career-limiting events or decisions—that both types of athletes will have significant career achievements during this time. Like a marathon, during the first half of a race, the legs are feeling fresh, and the finish line is a distant thought. The second half of the career marathon, however, is where it really counts to win. If we observe the vast majority of either type of athlete, the more significant career achievements often take place in this second half, when we have a greater appreciation for the upper quadrant of the CAHPT and continued accountability to purpose. This portion of the "race" is about pacing for the long road ahead while also remaining focused on hitting the important goals to ensure success beyond the finish line, giving consideration to your personal legacy. After transitioning into the second half, everyone has to dig a bit deeper, stay that much more focused, and be resilient to the challenges that may arise in order to maintain the pace in the face of fatigue. The second half is where it counts.

An easy way to illustrate this is to compare two different people with similar qualifications and capabilities and who graduated university at the same time. If we compare their careers after forty-two years (a symbolic unit keeping to our marathon example), when they retire from their profession, one will have had a very successful career with great accomplishments and the other a more moderate career with fewer accomplishments. Most people with similar qualifications and capabilities achieve similar success and results in the first half of their careers because the focus is on par: graduation, internship, job experience, career development, and climbing the corporate ladder. But there is an extraordinary difference in success in the second half because while they have the potential for greater success, it is more difficult for people to succeed. Progress is steady in the first half as they grow in their career and then that growth suddenly slows down significantly in the second half. Why do most people find it easy to succeed in the first half of their careers and difficult to succeed in the latter half, even though they may have more satisfaction? There are two factors that explain this:

1. The nature of the CAHPT components
2. The impact of your team

THE NATURE OF THE CAHPT COMPONENTS

The nature of each of the CAHPT components in the first half of a person's career differs from the second half and is responsible for major growth. In the early phase of their careers, people start at the bottom of the trapezium, which has a wider base, while in the later stage of their careers, they are in the narrower part of the trapezium. In the first half, the base focuses on the commitment to individual values, followed by the physical capacity to focus on the knowledge necessary to maximize career opportunities. The operating principle is that if you are good enough to do the job at these levels, then the next level should not be a problem and opportunity is a natural progression. However, in the second half, similar to the typical hierarchical organizational chart, the CAHPT narrows. Here, you begin to fine-tune skills relating to emotional intelligence and mental stamina. At this point, you may have achieved the ranks of management, and the difference between these fine-tuned skills is more apparent. Suddenly, it is not just a matter of how good you are, but how good you are relative to others and your effectiveness as a leader. At an organizational level, those who succeed in a management position may take on the role for a longer tenure. With longer tenures at the senior level, the next opportunity perhaps does not come immediately. Hence, opportunity becomes a constraint in the second half of your career and breaking out might even require achieving a major milestone and helping the organization toward a major achievement. The ultimate layer is purpose, which is the engine driving every other element. While the wisdom of purpose may feel like a late-second-half trait, can it assist in capturing the fewer opportunities in order to win where it matters?

THE IMPACT OF YOUR TEAM

The presence and effect of coaches and the people in your circle who help you succeed in each layer of the CAHPT is much higher in the first half than in the second half of your career marathon. At the beginning of your careers, the results you produce are not just a function of your capabilities, drive, and what you do, but also a function of the very active input you receive from your coaches and your team. In an optimal setting, during the first half, coaches and the organizational systems would be designed around the necessary training to address an individual's weaknesses. It would not impede your career progress, as it would help to accelerate and maximize the potential of an individual.

However, in the second half, the supervision (or lack thereof) you receive from your coaches in any organization is limited. Results are produced based on your own capabilities, and impact can be measured far more accurately. With a lack of coaching, there is no support to strengthen any weaknesses, which are often exposed at this stage. Inevitably, you struggle more often in the second half of your career due to these reasons. Ultimately, coach or no coach, there has to be an internal striving voice catapulting you forward.

TAKING CUES

That being said, you have to be open to the input from others, even if the message is not initially coming in loud and clear. You cannot let your focus on your own gains cloud your understanding of the needs of your team because there is no denying the benefits of having external accountability from others.

I learned this lesson the hard way. In 2012, Jason Sarai, a budding entrepreneur, and I met through mutual friends and we immediately hit it off. It was around what our wives would refer to as the third date of our bromance, when I invited him for a run. Unbeknownst to me, Jason was not expecting my call, as I had cancelled our run together earlier. He had settled down for dinner already, but being the good sport that he was, he agreed to meet up afterward. He showed up at my door as the sun was going down and the dusk was settling in.

"How long are you thinking?" he asked.

"Something short, probably eight to ten and we can see how it goes."

"Okay, sounds good!"

Having played soccer at the NCAA level and being physically active, Jason was not new to the world of elite training. In the first eight minutes of our run, he wanted to clarify the distance I was targeting and this is where we hit our first snag. "Eight to ten" for Jason meant kilometres; for me, it meant miles. To complicate things further, I did not know the actual distance because I had just planned a route using Google Maps that seemed interesting and was a contiguous loop back to the start. We were already almost a mile in at this point and I was so caught up in my own training that I nonchalantly dismissed his surprised reaction, figuring he would adapt. Keep in mind that Jason was running on a full stomach and was now completing approximately six kilometres more than he had initially anticipated.

I kept with talking pace and was interested in getting to know Jason on a deeper level. In my experience, the time spent cycling and running next to someone forges kinship. Add a bit of harsh weather and, suddenly, that kinship is intensified. Right on cue, it started to rain heavily. At this point, I was beginning to feel bad about the whole thing, but I was also determined to make this a positive experience for the both of us. I personally love running in the rain, but my friend, who was not training for anything and thought we were only going to run a maximum of ten kilometres, suddenly found himself in the midst of a sixteen kilometre-ish nightmare in a torrential downpour in the dark. We were about one-third of the way through the run when I glimpsed Jason starting to slow down and grimace. We still had about fifty minutes left in the run.

When we had about 5.5 miles left to go, I picked up the pace, leaving Jason struggling in the pouring rain. Since we had just met and I did not know where he lived, I was clueless that we had already run past his house. Wanting to keep the pace at a speed I preferred, I would run ahead and then run back, or sprint ahead and then wait longer at a set of lights so

Jason could catch up. I thought I was being encouraging but looking back, it was so obnoxious. Almost two hours later, we made it all the way home, with Jason knowing full well he could have called it quits and turned down the driveway of his home a half hour before.

In the end, we completed a distance of 11.2 miles. Although we finished a remarkable run and our friendship continued to blossom after this night, I am not terribly fond of my performance that day. I pushed my agenda onto someone else and I clearly did not listen to the signals Jason was giving when they were so evident. Above all, I failed to address a miscommunication when it arose and chose to ignore it because I believed my needs took precedent. A few years ago, we were laughing about this early experience and he admitted that at one point when I was darting ahead, he needed to stop and throw up his dinner. I felt truly awful. If Jason was not a genuinely good-natured, forgiving person, this would have been the end of the friendship before it even began.

I was hyper-focused with my own training plan and I could not see the forest for the trees. I forgot why I invited my friend to run with me in the first place, hoping to build a relationship with someone I found interesting and I also failed to realize what his own comfort level was because I refused to listen to the cues. I have learned so much from that incident and knowing Jason's personality and both of our friendly competitiveness, I now take a moment to think about what distance I want to suggest before we partake in any riding or running together. Like me, Jason is not a fan of mediocrity. While we both remain spontaneous in our workout regimen, we have both learned to check in with each other for feedback.

In hindsight, I was an ineffective cheerleader and blatantly misled Jason. Forcing people to come along for the ride is a rookie mistake for athletes and

Corporate Athletes alike. Reading the room in order to check in with both yourself and others is a considerate skill to exercise because it allows you to gauge the pace and comfort level of the team. There is so much information to be gleaned from seeing how everyone is doing, and if you are out of pace with your teammates, this gives you time and space to speed up or slow down. More importantly, if you are refusing to take in valuable feedback, conscious or subconscious resentment will start to build. Sure, the crowning achievement should be kept in your sights, but it is equally imperative to step back and really listen to the people around you. Is this a positive experience for them? Do they feel safe? They may be expressing vital information that you may miss when you are wrapped up in your tunnel vision. Taking a second to assess if everyone is still on board and feeling strong can only help. These regular check-ins are crucial to maintaining an overall healthy team.

BRIDGING THE PLATEAU

The key to transitioning from the initial ascent, through the inevitable plateau, and into the second portion of your career is passion. Why are you doing what you are doing? What do you love about your work? Capturing this energy will be what sustains you, allowing you to maintain pace and keep your sights on long-term goals.

While it is unrealistic to believe that your career will be exciting every single day or that you will love doing what you are doing every single minute, reminding yourself of your purpose and the passion that you derive from it will be the force that will keep you forging ahead. There will be many boring daily tasks that you will have to complete, and it is easy to lose sight of the big picture when you are embroiled in the daily details. But finding the energy, the drive, and the inspiration from your goals will fuel your pace and help you push through the plateau, transitioning into the second part of your career.

BREAKAWAY

EASIER, NOT HARDER

I am lucky I did not grow up in the flatlands. My dad anchored my running with hill repeats from an early age and the challenge of uphill running singlehandedly improved my technique and overall tolerance for distance running. Assuming you might be from the flatlands, my best advice to you is to learn to run uphill. By sharpening that skill, you too will improve your overall running.

The next time you run, count the times your right foot strikes the ground in twenty seconds. Multiply by three and you will have your stride rate per minute (one stride equals two steps, so your steps per minute will be twice your stride rate). Then speed up until you are running at eighty-five to ninety strides per minute.

The most common mistake runners make is overstriding, taking slow, big steps, reaching far forward with the lead foot and landing on the heel. This results in more time on the ground with more force on the landing, creating more impact on the joints. Training at a stride rate of eighty-five to ninety is the quickest way to correct this problem because short, light, quick steps will minimize impact force and will keep you running longer and safer. It will also make you a more efficient runner. Studies have shown that nearly all elite runners competing anything between 3,000 metres and marathon distances are running at eighty-five to ninety-plus stride rates. My coaches used to train running with a metronome. Nowadays, there are plenty of websites that list music by BPM (beats per minute)—either 90 or 180 BPM songs will do the trick.

Most of us have heard the story of the tortoise and the hare. It is the story of the race between the swift, agile hare and the slow, lumbering tortoise. The rabbit starts at a rapid pace and then takes a break, while the lumbering tortoise carries on without a break and finally wins the race. The moral of the story? Slow and steady wins the race. Today, when I see young entrepreneurs chasing careers, it often looks like the story of the hare. They get out of the blocks in a hurry, create a frenetic pace of career growth in the initial stages and then lose steam when it matters. Even so, I do not advocate "slow and steady wins the race" for careers, nor do I advise people to just focus on the finish line. I do, however, have an adaptation of that moral for the Corporate Athlete: Focus on a steady tempo and win when it matters. Remember that a constant cadence is key to reaching your goals, as opposed to taking large leaps or sputtering starts and stops. There is a saying that "time kills deals," meaning the art of closing business deals is undeniably linked to the cadence and the energy you bring to it. The same is true of the career of the Corporate Athlete or that of a runner. You do not have to win all the time. In fact, the more attention you pay to pace, the more success you will likely achieve in the long run.

You will likely have to revisit this passion, so the energy stays fresh to keep you motivated. So long as you find excitement in your purpose, it will be easier to get through the challenging times.

If you need more motivation, revisit the ideas of the Corporate Athlete High-Performance Trapezium and remember how capable, influential, and empowered you are when you are performing at your best.

AWARENESS

In one of my early jobs, I was the head of corporate communications for a biotech company, and the job involved travelling with the CEO, participating in investor roadshows, and managing grueling schedules. I was often trying to maximize the days in each city, scheduling as many as eight one-on-one investor meetings in a day in addition to larger presentations over breakfast, lunch or dinner. Our CEO would have to be alert and precise as he delivered the company presentation and whenever he concluded a meeting, he would often ask me, "How am I doing?" Effective leaders are comfortable asking that question and they expect honest answers. Sports teams review game footage in the same way, looking for cues and studying the plays to improve their effectiveness as players and as a team. Leaders understand that to make sound decisions, they need to know what is going on. They require honest and accurate feedback from colleagues, employees, vendors, customers, potential clients, the marketplace, and the community at large.

In 2011, when I first became a CEO, the roles were reversed, and I appreciated the importance of asking my teammates, "How am I doing?" Usually, the responses were positive and expected, but there was one incident where I received a very tough evaluation after delivering a presentation in a company-wide meeting. Immediately after, one of the employees on our team sent me an email, asserting that I had given the worst presentation, one deserving of a failing grade due to the disorganization and lack of clarity. What was worse, this feedback was from one of my direct reports, a colleague from the C-suite—the team composed of the chief financial officer, chief medical officer, chief scientific officer, and chief operating officer. I walked over to his office and mustered the courage to let him continue his barrage of detailed criticisms in person. This was a defining moment for me as a leader—I needed the feedback. We had developed a culture of people being able to deliver their perspectives without fear of retaliation. This particular opportunity created more trust and a deeper relationship between us.

As a young CEO, it was a humbling, frightening experience to be on the receiving end of an extremely visceral response from someone about a straightforward corporate update. Having a follow-up conversation helped clarify the situation and it turned out to be one of the most productive and positive things I have ever experienced. A year later, our company was invited to ring the opening bell at Nasdaq and after the ceremony, I made some closing remarks to the staff that was present. The same colleague came up to me afterward, offering his congratulations, not for ringing the Nasdaq opening bell, but for one of the most positive speeches I had ever delivered. This type of honest feedback has been invaluable in helping me become more effective as a leader and the positive impact that it had on the company and workplace culture was considerable.

BREAKAWAY

MANAGING ENERGY AND BUILDING STAMINA

Regular running is satisfying, but if you are the competitive type, even greater satisfaction lies in running faster and longer. Progress can be a wonderful motivator. If you want to improve as a runner, you can (and should) do supplemental training, which involves strength training, flexibility, and technique work.

The simplest way to improve is to run for time and increase the intensity over time. In short, set a goal to run for thirty or sixty minutes—something longer than you had previously accomplished. After you have been running for thirty to forty-five minutes at least three times a week, you can then increase that distance goal or start increasing your speed and intensity. Based on my experience, after six to eight weeks, you are ready to start running occasionally at 85 to 90 percent of your physical capacity. Build to where you can maintain that threshold level for five minutes. Then take one minute of easy running to give your body time to recover and repeat. As you progress, increase the number of the intervals and their length while maintaining a 5:1 ratio between work and rest. For example, you would complete ten-minute intervals of hard running followed by two-minute breaks, or fifteen minutes of hard running followed by three minutes of rest, and so on. After four to six weeks, you will be able to maintain this effort level for forty-five to fifty minutes, and you will not only be faster, but also build stamina.

Over time, you will build efficiency in running and your body will adapt to running with higher intensity for longer periods of time. A Corporate Athlete can benefit from adapting similar principles. Most people equate working hard with working longer hours, and the challenge with that is that time is a finite resource. Energy is a different story, and it can be improved with practice and sustained by stamina. Defined in physics as the capacity to work, energy comes from several sources in human beings: the body, emotions, and mind-spirit. In each, energy can be systematically expanded and regularly renewed by establishing specific rituals or behaviours that are intentionally practiced and precisely scheduled, with the goal of making them automatic. Similar to running, with practice, you can improve on each of the different sources to maintain energy and call on your stamina when you need it most.

The body is focused on physical energy. Not surprisingly, inadequate nutrition, exercise, sleep, and rest diminish people's basic energy levels as well as their ability to manage their emotions and focus their attention. Emotional energy is focused on taking control of your emotions. Unfortunately, without intermittent recovery, we are not physiologically capable of sustaining positive emotions for long periods. Tony Schwartz notes, "Confronted with relentless demands and unexpected challenges, people tend to slip into negative emotions—the fight-or-flight mode—often multiple times in a day."[18] The third energy source, the mind-spirit or the energy of meaning and purpose, is important to consider. People tap into the energy of the human

18 Tony Schwartz and Catherine McCarthy, "Manage Your Energy, Not Your Time," *Harvard Business Review*, October 3, 2019, hbr.org/2007/10/manage-your-energy-not-your-time (accessed on October 25, 2019).

spirit when their everyday work and activities are consistent with what they value most and with what gives them a sense of meaning and purpose. If the work they are doing really matters to them, they typically feel more positive energy, focus better, and demonstrate greater perseverance and resilience.

Regrettably, the high demands and the fast pace of corporate life do not leave much time to devote attention to these sources of energy. The solution is to bring awareness to the use and conservation of energy, as they are paramount to increasing endurance in your career and in your athletic endeavours.

A leader has to constantly look for ways to grow and improve. You have to know how you are doing from the people on your team. I am amazed by the number of people in senior leadership positions who do not know what is really going on in their organizations. They may know what the numbers tell them, they may read the reports and hear the updates from the department at the leadership meetings, but they do not have any firsthand knowledge about what is really going on at the heart of the business. The tricky thing about being out of touch is believing morale is good when it is not, and your colleagues are happy when they are not.

The same principle applies outside of the office—as a Corporate Athlete, you *should* know what your stakeholders and customers are saying about you. Investors, stakeholders, and customers define your brand, and they are your most valuable advertisers.

In their book *The One Minute Manager*, Kenneth Blanchard and Spencer Johnson encourage readers to become invested in their employees, to walk around and speak with them to get a real-time sense of operations on the ground[19]. Connect with your people. Get your hands dirty. Keep your ears open. There are tremendous payoffs for creating open communication and feedback with your employees, colleagues, vendors, investors, partners, customers, and community. It takes time to build a foundation of trust. Put in the time to review the "game footage." Knowing what is really going on is essential not only to success, but also to your very survival as a business and as an effective Corporate Athlete.

19 Kenneth Blanchard and Spencer Johnson, *The One Minute Manager* (United States: William Morrow and Co, 1982).

EXPECTING FAILURE

When I ran the Penticton Ironman in 2011, my childhood friend Winnie agreed to train together for the competition, even though she was not going to participate since she had completed the same course in 2008. We have been friends since we were pre-teens, swimming together in Vancouver, and she is a phenomenal athlete herself, having also competed in the New York Marathon and Boston Marathon. I remember when we would finish our economics class at SFU and then hit the university track together to do negative splits until we made each other cry or vomit. Her commitment to excellence is staggering. Personally, I had never ridden 112 miles before and she was adamant that I become familiar with the course in Penticton, taking me on a practice training ride through the bicycle portion one weekend. Ever the planner, Winnie is always expecting the unexpected, which enables her to think and create solutions for unforeseen circumstances.

The discipline that is required to tackle the cycling distance can be demanding, but she showed me I could alleviate a lot of the fear and diminish any out-of-the-blue challenges by preparing properly. She drafted a schedule for us to eat every thirty minutes and marked out the times we were to hydrate. She knew the course in detail, prepping me for the multiple ascents, giving advice about wearing the right gear or pointers about taking advantage of the descents to rest my legs and fuel up. We reviewed and practiced the route that weekend, both physically and mentally, her training teaching me the value of knowing what to expect in terms of the terrain as well as the importance of building and maintaining the stamina to travel it. Her attention to detail was exactly what I needed. She even mapped out the pivotal points in the course that could be helpful for me to take advantage of during race day. Her aim was for me to build resilience in order to reach certain milestones along the course and above all, to keep my head in the game. By doing everything she suggested, I had the confidence I needed to reach the finish line on race day.

On the last day of this practice run, we completed the full 112-mile cycling course. As we dismounted the bicycles, she motioned at me to lace up my running shoes to start the running portion, something we had not previously discussed. We immediately went on a mini eight-mile run so I could get a feel for the transition from bicycle to foot. As we were nearing the end of the eight miles, she led us away from our determined finish line, adding another mile to our practice loop. Having completed an Ironman on this very course, Winnie used her experience in the competition as the basis for our training. She understood what it felt like to be close to the finish line, only to be led away from it. She wanted to make sure I was prepared for all these little surprises that could derail me on race day. I never forgot it. I experienced the exact same feeling, of being so close to the end, but also being so far away from it at the same time. It was humbling. When things got rough toward the end of the race, I was able to replay that feeling in my mind. I was able to push through to the end. Her training enabled me to tap into my resilience so I could keep going.

Winnie was the first person who taught me how to see past the finish line and to expect failure. She is not a pessimist; she is a realist. She understands the value of not only pushing yourself to the limit, but also thinking of all the circumstances that could go wrong. When you expect failure, you are not conceding to it. Rather, you are forming your contingency plans, so you know exactly what to do when you encounter failure—and you most definitely will, I can guarantee it. Instead of spending all your energy trying to avoid it, what can you do to make yourself confident in the face of it? How can you give yourself permission to fail and fail well?

BREAKAWAY

80/20 TRAINING APPROACH

New Zealand coach Arthur Lydiard revolutionized the world running scene with his "mostly slow" coaching methods in the 1960s. In 1945, Lydiard conceived the idea that the key to maximum running fitness was a lot of slow running.[20] He believed endurance was the true limitation in running, and training programs should emphasize endurance building, his mantra being the "secret to running faster was to run farther." Through his own trial and error in training, Lydiard found that speed work helped the most when it was "sprinkled lightly" on top of a large foundation of slow running. He pioneered the 80/20 training approach, which is widely used today, where the emphasis placed on lower intensity runs is believed to create greater success.

Following the impressive success of Lydiard's athletes at the 1960 Olympic Games in Rome, his low-intensity, high-volume approach began to spread globally.[21] In the modern era, all distance world records have been set by athletes trained with an 80/20 approach, including the current distance domination of the East Africans, including the Kenyans and Ethiopians. The difference between runners who realize their full running potential and those that do not is the amount of slow running each person undertakes.

20 Arthur Lydiard and Garth Gilmor, *Running with Lydiard* (2nd ed.) (United States: Meyer & Meyer Sport, 2000).
21 Ibid.

It is difficult for many runners to make peace with the concept that if they want to run faster, they need to slow down some of their training sessions because it feels so contradictory. In *80/20 Running: Run Stronger and Race Faster by Training Slower*, author Matt Fitzgerald outlines that recent studies of the training practices of the world's leading runners reveal they spend an average 80 percent of their total training below the ventilatory threshold. The ventilatory threshold pace is slow enough that a runner can hold a conversation. In well-trained runners, the ventilatory threshold falls between 77 and 79 percent of maximum heart rate. In other words, for every hard run, the elite distance runner will run four easy runs. By contrast, the recreational runner tends to run one easy run for every hard run. The other 20 percent of training time is spent at high intensity—that is, above the respiratory compensation threshold (the point where hyperventilation or rapid, deep breathing occurs).

Fitzgerald cites that new research suggests recreationally competitive runners improve most rapidly when they run more slowly in training. The good news is, unless you are an elite runner, it is almost certain you are doing less than 80 percent of your training at low intensity, and you can improve simply by slowing down.

People often scoff at the idea of having a Plan B, as if that means you do not have your full weight behind your ultimate goal. But I believe that Plan B (and C, and D, and E) will be the safety net that will help you not only recover faster in the face of setbacks, but also provide the blueprint to help you prepare for the worst. When you are ready for any challenge, setback or curveball, you are no longer on edge, but relaxed. This confidence is more beneficial than actively trying to avoid failure, which can be paralyzing. Do not be afraid of your backup plans—use them to build confidence and resilience in the face of adversity. Expecting failure can actually help you greatly in the long run.

INTENSITY BLINDNESS

There is a contradictory training regimen that professional athletes undertake that enables them to extract the most out of their training time and athletic abilities in order to reap the biggest improvements: moderation over time. On the surface, it appears antithetical. But if athletes consistently train at a lower intensity and limit the amount of maximum exertion effort, they will show marked improvement over time. The more they pace themselves, the less chance they have of overdoing it, burning out or risking injury and ending their careers altogether. Going hard and going fast does not necessarily get you there in record time.

The desire to want to do too much too soon is called intensity blindness, and it has applicability that extends beyond training into business. In the world of entrepreneurship, it can distract you from your purpose, derailing your work. I have witnessed many projects fail because of it.

In numerous biotech companies I have been involved with, most often, management is proud of the technology because it has a broad range of applicability to multiple different types of unmet diseases. I have heard many CEOs present to their board of directors, including myself, with an intent to demonstrate a disruptive solution across several applications by establishing multiple clinical trials as opposed to a very focused plan around the single stronger probability of success. In biotech, this is the equivalent of putting all your eggs in one basket. As well-intentioned as the plan is, it is merely a way to demonstrate the drug's potential capability to our stakeholders and investors. I admit when I first fell into this trap of maximizing potential, it sounded great to me too, letting this massive undertaking really show what we could do rather than just relying on one trial. But more often, we fail to consider how such a venture might exhaust our resources and bandwidth and take away from our focus. Even stakeholders and investors have a tough time getting past the primary program. Here is where the concept of calibration comes in: Rather than choosing to focus between one thing versus many things, perhaps you can split the difference. Calibration is an important part of moderation. It helps you find the happy medium that is often required in business because very rarely is anything all or nothing. In my experience in biotechnology, it is imperative we find a drug that can solve a problem; it is also equally imperative we look for other ways this drug can also apply to other diseases. We need to understand that we cannot find a million solutions at once and also, we cannot be so shortsighted as to not realize that this one drug might be applicable to other conditions, saving more lives. A little from column A and a little from column B—this is the golden ticket to success.

As an entrepreneur, we can make far more impact by making sure we do one thing very effectively rather than failing at multiple things at the same time. But intensity blindness is common among us because we often feel it is what our stakeholders are looking for. It is tempting to want to try to impress everybody and it is easy to believe you always need to do more when the old saying is true: less really is more.

Apple's launch of products like the iPhone is a perfect example of this. They have a yearly cycle of one major innovative idea. Yet, it is so disruptive it changes the course of the industry every single time. Tesla operates in the same way. The company did not launch five cars at once, but rather it launched one, allowing the public time to be wowed by their product before giving them a new variation. Tech companies are notorious for doing this very well, presenting minimal viable product for launch and then coming out with fresh features to augment the original product.

In life sciences, if your drug works in multiple applications, it is common to want to go after unmet medical needs in many different areas. Why would you want to thwart your pace to make these treatments available? After all, it makes sense from an altruistic perspective because you are trying to solve multiple diseases and giving patients options they would not otherwise have. But at the same time, if it is so early that you do not have the confidence in the efficacy and you are going forward based on risk, then you are also setting yourself up for failure and potentially impacting your lead product or clinical trial's chances of success as well. By trying to go after everything, you forfeit your reliable moderate pace.

Why would a runner naturally choose to habitually train faster than what is comfortable or required? In *80/20 Running*, Fitzgerald cites that this is due to the task-oriented nature of runners. When a runner has a "job to do," they want to get it done. Most runners think in terms of covering a distance as opposed to running for time, and the fastest way to get a distance-based task completed (e.g., a 5K run) is to push hard and try to run it fast. The same could be said about entrepreneurs, as most are thinking about the endgame and quickly spinning to do as much as possible toward the commercialization of a product and fail to take stock of resources. We want to build resilient Corporate Athletes who can extend their endurance and think past the finish line.

This is not to say that you do not want to be prepared should your original plan fall through. In drug development, it is common to have "multiple shots on goal," meaning if you have a failure in your first target, you want to hit the ground running with the second to minimize any setback and value in the business and avoid losing confidence from your different stakeholders. It requires a delicate balance, and every situation is unique, but in my experience, focusing your energy on the thing that matters the most equals a better chance of success.

RETHINK, REASSESS, REVIEW

If you feel yourself starting to take on too much, there are ways to rein yourself in. Start by asking yourself what is really required for success. Does success require ten clinical trials or will one suffice? Does it mean launching five new products simultaneously or could one successful launch lead to more down the road? When my company was developing its melanoma treatment, our initial trials showed a higher response rate than the current standard of care for that particular patient population. While we felt fortunate because of those results, we knew we needed to show its impact on a larger scale. That gave us the confidence to launch a larger clinical trial to show an even stronger statistical significance as well as a higher bar for risk probability. We had set criteria that we required for the success of the outcome.

Ask yourself, "What is my primary objective?" In biotechnology, the answer is always safety and efficacy. You want to make sure your drug is safe and shows relief of the symptoms and resolution of the disease. In the best case, you are showing it to be statistically significant compared to an alternative or to the existing standard of care.

Ask yourself what you are doing to protect yourself against any failure because if you have not planned for it, you have already failed. Furthermore, when you do not make the right decision, you should be prepared to accept the consequences. You cannot divert blame to others; it is important to own up to your actions and, in my experience, that does not have to be a negative experience. By taking responsibility, you are reinforcing the accountability you feel toward yourself. You can handle the situation and move on and there is power in that.

I also encourage you to analyze how successful your personal and corporate responsibility strategy is in engaging stakeholders and creating goodwill. You can do this in myriad ways, but for me, YELL has given me the structure I need in order to be able to balance what I need to do to earn a living as well as participate in matters of education reform and entrepreneurship. Think about your own actions: If they are leaving a negative wake, then stop doing them. If they are leaving a positive wake, then recognize how those actions are helping you live your purpose.

BREAKAWAY

MASTER CLASS IN TRAINING STRESS

"Never give in, never give in, never, never, never, never—in nothing, great or small, large or petty—never give in except to convictions of honour and good sense. Never yield to force; never yield to the apparently overwhelming might of the enemy."
—Winston Churchill

San Diego is home to one of only two Navy amphibious training bases in the United States. The Naval Amphibious Base (NAB) Coronado provides a shore base for the operations, training, and support of naval amphibious units on the West Coast. Likewise, current SEAL Teams are organized into two groups: Naval Special Warfare Group One (West Coast) and Naval Special Warfare Group Two (East Coast), both of which come under the command of Naval Special Warfare Command at NAB in Coronado, California.

In 2009, I met Steve Nave in San Diego. He served twenty-three years in the U.S. Navy, with eighteen years as a Navy SEAL. He was also a Navy SEAL instructor in the Naval Special Warfare Physical Training Center in Coronado, California, and trained the men in BUD/S (Basic Underwater Demolition/SEAL) who go on to become SEALs. Steve understood what it meant to take people to the edge and beyond, testing their physical strength and endurance as well as increasing their mental stamina.

Every summer, Steve would host a series of Navy SEAL training challenges at La Jolla Shores Beach. This was nothing close to the rigorous years of training that the incredible men and women of the US Navy SEALs endure, but served as a condensed version of BUD/S. Over the course of six to eight hours, as a civilian, you can experience a portion of what it feels like to go through a strict twenty-six-week training regimen. Sign me up!

7:00 a.m.: Meet-up point at La Jolla Shores Beach. Everyone was pre-briefed to be on time and ready to go, wearing a T-shirt or long-sleeves, shorts, and runners.

7:01 a.m.: The person I recall being a super nice guy has now transformed into Drill Sergeant Nave and is ready to administer stress and training like no one has ever experienced before.

Over the next six hours, I felt like my body and mind were not necessarily connected and my brain's reaction was to instinctively enter instantaneous survival mode. Steve was the epitome of what a drill sergeant would look like in the movies: bald, very muscular, incessantly shouting. He had the physique to boot—very muscular with pronounced calves and arms characteristic of a Navy SEAL. The one thing about being a civilian doing this type of training is the reassurance that beneath all of the bravado lies a deep understanding of sports science—at least that was what I was convincing myself. *He won't let me die, right?*

7:02 a.m.: Steve instructed everyone to assume a plank position and count off push-ups together. "Yes, today is going to hurt and your body will be in a bit of shock. But for SEALs to become SEALs, they have to be ready for any situation and deal with the stress inflicted on the body." At this point, we were nearing twenty push-ups and then were instructed to remain in a plank position. "You will not be able to adapt to become stronger without experiencing some form of stress, so let's do it again, sound off, twenty more." Everyone was grunting in agony. Many of my fellow mates who thought this was a good idea just like I did were seriously rethinking what they signed up for. Steve turned to one of them and barked, "What are those? Do not quit! Keep your body in one long line, bend your arms and lower yourself as close to the sand as you can. Eighteen! Nineteen! Twenty! Onto your back!" We proceeded to do the same for military standard sit-ups, knees bent, feet remaining on the ground, fingers clasped behind our heads, coming up all the way until our backs broke. Over the next six hours, I did more push-ups, sit-ups, and burpees than I could count along with running, sprinting, swimming, training in surf, carrying a buddy, and other team-based training that involved staying locked in arms while withstanding the surf or carrying telephone pole-sized logs over a specific distance. Some would refer to this as torture, but I had a phenomenal experience. Over the course of the day, Steve broke up his drill sergeant tone to provide detailed instructions on how to carry out specific exercises with efficiency and avoid injuries as well as practical explanations on how we were taxing the body's cardiorespiratory system. He explained that the rationale behind physical training instructors in the Navy making recruits do extra push-ups, run another mile or carry more weight in their backpack was not to be cruel but rather to produce physical adaptations so they could improve.

He was right. This was an idea broadcast to the world of sport in the 1960s by Fred Wilt, the editor of *Track and Technique*, who published the article "Stress and Training."

The idea of physical adaptation or using stress for training sounds very simple, but it is amazing how many people still fail to grasp this. To improve, you must expose your body to a specific stimulus to the quality or intensity of which it is not adapted. There is no magic pill, quick fix or shortcut. You are going to have to sweat. You are going to have to work. At times, you might not like disrupting your comfortable state of homeostasis. Needless to say, over the six hours at La Jolla Shores, the state I was in was anything but comfortable. I was covered in sand, which reached crevices in my body that I never knew existed, and I had to succumb to the sand chafing. I was drenched, with my clothes and shoes weighing me down, my senses numb from the cold, my body aching. Every part of my body was in pain, but I kept telling myself that this meant every part of my body would adapt.

For the last part of the experience, we were told to drink, eat, and get dry. Note the word "told," rather than suggested. This is because the Navy SEALs understand recovery is critical and at times, short.

Yes, stress and stimuli are needed but on the flip side, there are people who run, jump, and lift as much as possible, as often as possible, every hour of every day. This too is destined to fail because of overtraining. Hans Selye describes different stages of stress, known as the general adaptation syndrome (GAS). When you understand the different stages of stress and how the body responds in these stages, it is easier to identify signs of chronic stress within yourself.

Everyone has those days when you feel tired, with low energy. When you have no strength to power through a weight lifting session and your burpees are sluggish, chances are, you have pushed the stress and stimuli too far. This likely means your homeostasis is not in balance, your internal environment is a mess, and your body is waving the white flag for you to surrender to the sofa. For this very reason, whether you are training to be a civilian Navy SEAL or before you even set foot inside a gym, you must have at least an understanding of the three theoretical stages of Selye's general adaptation syndrome:

- **Alarm/Shock Phase:** Upon encountering a stressor, the body reacts with a "fight-or-flight" response and the sympathetic nervous system is activated. Hormones such as cortisol and adrenaline are released into the bloodstream to meet the threat or danger. The body's resources are now mobilized.

- **Adaptation/Resistance Phase:** Blood glucose levels remain high, cortisol and adrenaline continue to circulate at elevated levels, but the outward appearance of the organism seems normal. There is an increase in heart rate, blood pressure, and breathing. Body remains on red alert.

- **Exhaustion Phase:** If the stressor continues beyond the body's capacity, the organism exhausts resources and becomes susceptible to disease and death.

In summary, the perfect adaptation is when you train to maximize the time spent in the Adaptation Phase, while avoiding any time spent in the Exhaustion Phase. According to Selye, "The loss of acquired adaptation during the stage of exhaustion is difficult to explain, but as a working hypothesis, it was assumed that every organism possesses a certain limited amount of 'adaptation energy," and once this is consumed, the performance of adaptive processes is no longer possible."[22]

---

22 Hans Selye, "Stress and the General Adaptation Syndrome," *British Medical Journal*, 1950, vol. 1, no. 4667, pp. 1383–1392, doi:10.1136/bmj.1.4667.1383.

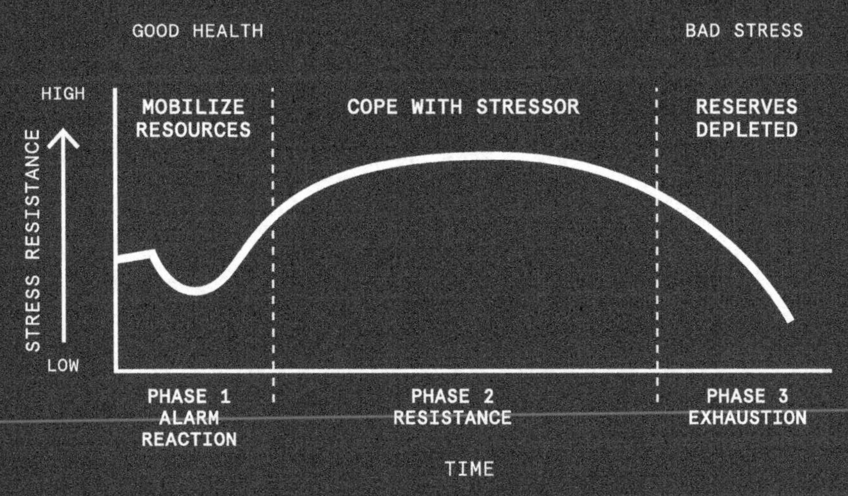

General Adaptation Syndrome

Of course, "adaptation energy" can be accurately measured with a few simple tests to monitor white blood cells, hormones, and the body's immune system. However, this may involve needles and I personally hate them. Also, relying on people in white coats to tell you how to train every day is not practical for most people. So instead, become your own expert and listen to your own body.

Although your program might specify a seven-mile run, do not feel you have to run that distance if you honestly believe your "adaptation energy" is low. Instead, work on mobility or something complimentary with lower intensity. Focus on drills and technique and light, active recovery. Or alternatively, just switch off and meditate. Bottom line—avoid the Exhaustion Phase and live to fight another day within the Adaptation Phase.

Sadly, Steve passed away in 2016, but his legacy lives on. His philosophy on life and his long-standing belief that you could achieve anything you wanted so long as you believed in it was not only contagious but also inspiring for athletes and nonathletes alike. He truly believed in mission accomplishment. His dedication to service and his ability to bring out the best in people will not be easily forgotten and I am humbled to have learned so much from him that day on La Jolla Shores Beach.

CHECK YOURSELF

One of my favourite things to do is to cross things off a list. The sense of satisfaction is a reminder that even the smallest completed goal amounts to something bigger. Checklists help you zero in on what matters most. If you are spending more time dwelling on what you need to get done than actually *doing* what needs to be done, you get nowhere. Having a checklist is also one way to protect yourself from failure. For example, if my wife sends me to the grocery store with a list of things she needs and I come home with all but one item, the whole effort was a failure. What is the point of having a checklist if you have no intention of following it? Checklists offer you an ideal opportunity to rethink, reassess, and review as you prioritize your goals and map out your actions accordingly.

I once had the privilege of sitting in the cockpit of a private plane on a cross-country flight. In speaking with the pilot, I learned everything they do is based on a checklist. They have a checklist for taking off. They have a checklist for when a certain light comes on. They have a protocol to follow for any possible situation, and if Plan A doesn't work, they always have a Plan B. The pilot explained that the system allows him to always know exactly what to expect. He had his own experience to rely on, of course, but the checklist was in place to ensure nothing foreseeable could go wrong. In running his checklist, he knew he was doing everything that was supposed to be done. Every Sunday, I sit down and write a list based on my goals for the coming week. Throughout the week, I refer to it constantly, making sure I am staying on track and remaining true to my purpose as I face the day-to-day challenges that come my way. To make an effective checklist, you first need to understand the biggest, hardest, and most important thing that needs to happen. I suggest starting with three of your top goals for the week with no more than ten subcomponents listed under each. The subcomponents are the tasks you need to do in order to ensure completion of that goal. Even if the three things on your list seem out of reach, you know the ten actions you need to do to support each are what will help you make progress. Ten is just a suggestion and you can do more or less depending on your needs. You should be able to remember your list without referring to it.

BEYOND THE TO-DO LIST

As a CEO, my process is even more elaborate than weekly checklists. My planning for the next year always starts in quarter three when we draft our five core milestones for any company with the components required for each listed underneath. Each objective is divided among my C-Suite, and every week I ask each member of the team to list their top three priorities pertaining to their specific objective, each one representing an extension of the things driving their performance metric. Underneath that, I ask them to list no more than ten things they needed to accomplish that week to make progress on the top three priorities, so we can see a percentage completion toward that top three. Every Monday, we circulate our lists among our management team. In turn, the managers do the same among their own teams and ask their direct reports for similar lists.

The process's entire purpose is to keep everyone aligned. Every time I or management reviews the checklists, we ask ourselves questions: "Is Bob focused on what he is supposed to be focusing on? In his top three, does it say something that does not pertain to our milestones?" I and the management team would like to know right away exactly why everyone is doing what they are doing. If they mention something that did not fit with our overall corporate milestones or corporate scorecard, we will simply ask for clarification or more perspective. It keeps the entire team focused on moving the needle on the five things we established in the beginning of the year as being critical to our success. It also eliminates any temptation to procrastinate, something all of us can inherently fall victim to. There really is no room for procrastination, especially in a field like life sciences where the primary objective is to help patients who have an unmet medical need and little or no other options available to them.

The lists also create a sense of urgency. It is easy to get bogged down in the day-to-day tasks and lose sight of the real reason you are doing the work that you do. We are not all necessarily thinking about the fact we are helping people suffering from stage four melanoma with less than a few months to live as we go about the business of opening a clinical site, finalizing a budget or attending another meeting. But these lists show the team how everyone's job matters, and even if some of the things on the lists might not sound terribly exciting, they are all necessary to help us reach our goals. By making the list, you are holding yourself accountable to getting it done—all of it—and by sharing it with the rest of the team, you become accountable to everyone else as well.

By making sure everyone stays focused on what matters most, this process becomes the key to our collective productivity. It shows us how our milestones are aligned to people's roles and responsibility, and how people's roles and responsibility are linked to their purpose-to-impact plan. Then, any success or achievement results in shared respect among everyone on the team as well as a shared happiness in getting it done.

Ultimately, this is also a great way to gauge progress because we can show stakeholders the specific steps we take to hit our milestones. It allows for total transparency between all parties, preventing burnout because everyone focuses on the things that matter most to them. Everything they do is aligned to their purpose, and all they need is a checklist holding them accountable to their plan.

In business, it is crucial to remember progress comes in all shapes and sizes. From C-Suite executives to any entry-level employee, we all contribute to a larger purpose, but only the individual can determine how small or large the contribution will be. Transformational leaders, entrepreneurs, innovators, and connectors who strive for a greater purpose and who look to be better than who they already are will be perfectly set up to raise the bar because operating in possibility creates endless opportunities for growth.

No matter how you opt to organize your priorities, make sure your process empowers you to simplify and conquer the complex problems. Tools presented in this book, like the purpose-to-impact plan or checklists, will assist you when you encounter challenges that will inevitably come your way as you work to make a difference in the world.

RAINY-DAY RUNNING

A lot of people avoid running in the rain, then find themselves unprepared when the sky opens up on race day. You simply need to override your discomfort and just do it. By this, I mean you will need to force yourself to get used to feeling uncomfortable, exposing yourself to as many conditions as possible so you are not caught off guard but better accustomed to the conditions for the future. You will have to find that intersection of motivation and resilience in order to push yourself past the discomfort and onward to new levels of performance.

Sometimes, that uncomfortable situation might be a conversation you have to have with a colleague or a stakeholder. It could be countering a decision that goes against your core beliefs and values. There are many difficult situations you may find yourself in and the easiest way to familiarize yourself with the hard challenges that may arise in your career is to roll up your sleeves and dive in. The more you practice existing in this uneasy space, the better acquainted you will become with unpredictability.

Jason and I may not have realized it at the time, but our run in the rain laid the groundwork for him to become better prepared for adversity in the future. He persevered through miscommunication, decided he was still up for the challenge, and navigated through unforgiving terrain, all in the name of friendship. Having experienced such unforeseen conditions that night, I am willing to bet that, today, he can run in the rain after dark, without any fear, and of course, with a full stomach. I would not expect anything less.

□

WORLD-CLASS CODA

- Highly motivated people are more apt to suffer from burnout syndrome due to their tendency to overcommit. The very characteristics that enable your success can also be your obstacles to success. Adopt a more moderate pace in order to conserve energy and create longevity in your career.

- Sometimes, doing less is actually doing more for your career. Moderation over time allows you to sustain your pace over the duration of your career, allowing you to take the time to appreciate your progress and achievements thus far.

- Having a contingency plan is not a sign of failure but, rather, sets you up for success because you have a plan for failure.

A safety net does not hold you back from success; rather, expecting failure gives you the confidence to face setbacks head-on, allowing you to cultivate resilience.

CATAPULT FORWARD

- How do you prevent yourself from experiencing burnout? What rituals do you undertake in order to ensure you have adequate work and play time?
- Why do you do the things that you love? How do you make work interesting, so you are invigorated by it as opposed to demotivated by it?
- What do you do in times of crisis? How do you cultivate resilience?

RISE UP

SECTION THREE

Communication is a two-way street. Speaking fearlessly with conviction and listening mindfully with patience are both learned and practiced skills. When you speak and listen actively and use conversation as a tool to encourage truth, you create space where honesty is favoured over deception as a constructive solution. Be brave, speak up, listen carefully, and be kind—you will thank yourself in the long run.

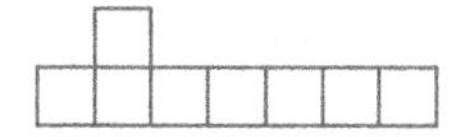

PRINCIPLE 7

BREVITY

"It takes two to speak the truth—
one to speak and the other to hear."
—HENRY DAVID THOREAU

It was a crowded room on the 86th floor of the Park Hyatt Shanghai and I was about to start an investor presentation. It was my first time in this city, and I had arrived late the night before from Delhi. Going on little sleep, I definitely had some good nervous energy. The sprawling rectangular room was made up of walls of crystal-clear windows with sweeping views of neighbouring skyscrapers and the Huangpu River down below.

Up until this moment, I had yet to admire the view, as I had arrived at the hotel under the cover of darkness the previous night. This morning was a whirlwind—I slept for two hours, fit in a short workout, and then got ready for several meetings and my presentation. A fireside chat was to follow, a format of discussion at investor conferences where an analyst or industry expert has a more candid talk with management that extends beyond the official corporate presentation. It was a very full day.

What no one knew was that I was coming from my grandfather's funeral in India. It was an unexpected, emotionally jarring time and coupled with the long travel time, showing up in a foreign city, and getting right back to work, it was more than challenging for me to step up to the proverbial plate. But this is what all my preparation was for—the repetition, the familiarity with the material, the practice of ingraining everything I know into everything I say and do is what sets me up for success when I am called to perform. Whatever I am going through is separate to whatever I need to accomplish professionally. Whatever turmoil I am experiencing cannot derail what I set out to do in Shanghai or anywhere else, nor should it.

As I was called up to present, there were the usual, expected technical issues that occur, so I was not thrown off. When those were resolved, I looked up to speak and from the corner of my eye, I saw the slides on the screen. They were all in Chinese characters. I did not expect this; they were always written in English. I gazed at the crowd of over 300 people, the biggest audience I had ever spoken in front of, their faces attentive, and I began the presentation. In my head, I was feverishly wondering if the audience understood English or if they were simply reading the slides in their native language or worse yet, they were comparing the two. I had no idea if what I was saying lined up with what they were reading, but I had no choice but to assume that it did and continue on. I had given this presentation many times. Prior to this particular event, I was frustrated by the repetition of it. I was unable to alter parts of it to make it more interesting, if not to myself, then to the audience, which is a very common practice when you have to stick to a script because of various corporate regulations. I hated to admit it, but I was getting bored by the predictability of the speech. Ironically, this repetition ended up saving me that day because I was able to rely on my sense memory to keep my composure and soldier on when I was completely caught off guard. Because of this speech-slide disconnect, I was forced to focus on primarily conversing with the audience, rather than solely reciting while relying on the slides alone, which enabled me to present in a more personal manner and with greater clarity because I was truly in the present moment. I relied on facial expressions as guideposts to surmount the language barrier. I was able to interpret that I was on the right track, excelling even, from the body language and collective energy in the room. I was able to maintain grace under fire.

The art of communication is indispensable as a Corporate Athlete. One of the primary ways we communicate is via the art of persuasion or selling. For example, getting to an outcome with a team is all in the delivery. It is a common belief that most leaders have a natural extroverted personality that makes them effective—I would argue it is not as natural as one may think, but rather more deliberate. If you look at what most leaders do all day on the job, no matter what their job title is, they are trying to sell something. If you are a coach, you are selling your players on a way to win; if you are the CEO, you are trying to influence your team to do something in a different way; if you are physician or a nurse, you are selling patients on new life habits or a therapy that can help with a devastating disease. Across all professions, you can look at what people do all day on the job and it is often very different from what appears in their actual job descriptions. As a CEO, I can admit that over 50 percent of my job is spent on something akin to sales, persuading, influencing, and asserting a degree of confidence. Outside of the work environment, this can easily extend into my role as a father while communicating with my kids, as I sell them on the idea of eating more vegetables. This is a large part of what we do for a living and when we speak to our teams about this and remind them that no matter what their job description may say, a part of their role and responsibility is going to be selling, persuading, and influencing, there is a natural apprehension. Most people feel they are not a "salesperson" or that they are not good at "sales." The truth is, most of us have an erroneous image and a very outdated notion of what sales are, when, in fact, we are all selling something during our many daily interactions, be it an idea, a point of view or a way of life.

Funnily enough, this presentation in Shanghai was all about authentic communication and was an opportunity to evaluate new business propositions and investors for OncoSec. My understanding of the material and my ability to deliver it with passion instigated several important investor introductions by the hosting bank. Sure, the timing was terrible because no one knew or understood the personal sacrifices I had made to be in that room, but I was able to tap into my muscle memory to not only get through it, but also perform to the best of my ability. Despite physical and emotional exhaustion, curveballs, and the lonely experience of being in a foreign city for twenty-four hours where I did not speak the language, I did not fall apart. I was able to build and rely on all the concepts I have discussed in the book thus far in order to excel. This is resilience.

SETBACKS

The corporate handbook unfortunately does not train you for dealing with delayed travel schedules or cancelled flights, nor does it provide tips for working all day and all night to close a deal. I admire the strict adherence to labour laws that most corporate handbooks abide by, but the reality is that they are built to protect the employer or employee and not designed to build resilience in the Corporate Athlete.

Throughout your career, you are going to be faced with countless unknowns and challenges that were not included in the first-week job orientation and you will need to adapt under the circumstances in order to get through them. The HR professional is going to expect you to work forty hours a week and probably turn a blind eye when you put in the extra hours that are technically not included in your annualized salary. Your boss is going to expect you to work sixty hours to *really* earn the salary you negotiated, and if you are the boss, you are never going to feel like you have done enough. If you want to prepare yourself as a Corporate Athlete and build resilience, you need to stay organized and ready for the demands.

1. **Curveballs:** As an entrepreneur, something is inevitably going to be thrown at you that you were not expecting. In 2019, I was on a company board of directors where the CEO was bringing forward a financing term sheet for a company where the deal seemed favourable and the right thing for value creation for the company. But the major shareholder, who

had a seat on the board of the company, did not feel the same as the rest of the members. In an ideal situation, the deal would have been completed in eight to twelve weeks, but what ensued was a lengthy proxy battle wherein the largest shareholder was motivated to block the deal and present inferior financing terms. This lasted over six months and became a massive drain of energy. Ultimately, what was originally presented was approved and supported by the remaining shareholders of the company, but it took much longer than expected. All to say: it will never happen like it does on paper. Timelines will be estimates and you will have to pivot as you go along in order to accommodate unforeseen circumstances. If you have thought through the different scenarios to the best of your ability and done everything to hit your objective, you need to remain focused and push your objective through to completion. Your ability to maintain a cool head and persevere when things change on a dime will be key to building resilience in an ever-changing entrepreneurial world.

2. **All-Nighters:** In college, an all-nighter is a rite of passage for students, especially during finals. As adults, we try to avoid them but sometimes we need to pull one in order to finish a work project or a deal. I have endured my fair share of all-nighters over the last twenty years, with each deal becoming that much more intense and with the feeling that there was so much more on the line. Pulling an all-nighter is not the most healthy or desirable thing in the world, even though it can produce positive feelings of euphoria. Moral support always helps, so make sure you have friends and coworkers at the

ready so you can get through the witching hour. The reality is, the more prepared you are, the less you will need to stay up through the night—it always pays to be on top of your own workload. Be prepared with a checklist of what needs to get done. Get some sleep when you can. In circumstances where things are out of your control, being prepared will still help you get through an all-nighter easier—at the very least, it will make things a little less painful. Just remember to breathe.

3. **Jetlag**: Before travel came to a screeching halt during the COVID-19 pandemic, between 2010 and 2019, I endured the heaviest travel period of my career, when I boarded a plane over 120 times every year. I have experienced delayed flights, detours, extreme weather conditions, broken all my personal bests from bed-to-boarding (thirty-five minutes!), and even missed a flight while waiting at the gate because I had my noise cancelling headphones on and lost track of time. When I was young, work travel sounded intriguing, but no one gave me the orientation on what was really involved in corporate/business travel. It is definitely not as glamourous as I once thought, and I am happy to be at a place now where I can choose to avoid it if needed. When you are tired and far away from home, the temptations to eat poorly are everywhere, especially in an airport or at restaurants. Comfort food is a real thing and when you are exhausted and miss home, you will want to eat all the foods that emotionally make you feel better. I suggest you try to stay disciplined with what you put into your body so you can stay on top of feeling your best. Fitting in a workout once a day while you are away will do wonders for your sleep when you are in different time zones. Most

importantly, try to stay in the current time zone, no matter how exhausted you are. Comparing your current time to the time it is back home is the kiss of death and will only confuse your circadian rhythm, making it that much more difficult to settle into a routine in your new city. Above all, make the most of your time in these new places while you are working. Take a beat to look up from your laptop to appreciate and find inspiration in your new surroundings. As hard it is to be on the road, make the most of experiences while travelling because when you step off the plane and walk through your own front door, you will have a new appreciation for home.

Adaptation is the name of the game when it comes to life, sport, and business. There will always be a wrench, a blown tire or a last-minute change that will upend your original plans. The key to dealing with these setbacks is to go with the flow, trusting that you will make the right decisions when you need to. It is so easy to give into the stress that comes when something unexpected happens, but the trick to navigating rough waters is knowing that over time, you will encounter enough disasters and will survive them. This confidence will carry you over any setbacks in the future and the experience will better inform you the next time something inevitably goes awry.

TALKING POINTS

Communication is a key principle for success and its importance in both sports and business cannot be understated. Effective communication is a way to help athletes, Corporate Athletes, and their respective coaches deepen their connections with each other, improving teamwork, accelerating skill learning, and enhancing performance overall.

One of the key advantages of effective communication in the sport environment is developing trust between coaches and athletes, which exerts a positive influence on the athletes.[23] This trust is established when coaches learn how to effectively express and transmit their knowledge, thoughts, and feelings to their athletes. In this way, coaches are more confident about the emotional impact of the content of their messages to the athletes. Effective communication is also cultivated when coaches pay attention to what is expressed by the athletes because communication is a two-way street that involves receiving messages as well as sending them. When coaches learn how to listen actively and respond, athletes can feel their coaches understand them. It lays the groundwork for expressing their goals, thoughts, and feelings successfully to their coaches.

Although there are countless studies that emphasize the importance of successful communication, little time is devoted to helping young leaders, athletes, coaches, and Corporate Athletes actually become better communicators. One particular study interviewed elite-level teenage athletes who suffered burnout and found it was not physical exhaustion that contributed

---

23 Simon Jenkins, "Short Book Review: Sport Psychology for Coaches." *International Journal of Sports Science & Coaching*, 2008, vol. 3, no. 2, 2008, pp. 291–292, doi:10.1260/174795408785100734.

to burnout, but poor communication[24]. The lack of communication with the coach created a perceived low level of personal control over the situation in which they trained, leaving athletes feeling stressed and unable to cope as they deemed their environment to be controlling and overly pressured. Another study conducted in Europe also noted the importance of managers being able to identify what motivates their employees[25]. This allows them to try to make work more engaging, resulting in employees feeling more motivated to create a better performance. For all this emphasized importance, how can we actively encourage people to work on and improve their communication?

There are multiple models of communication that are beneficial to the world-class Corporate Athlete and will help them navigate their path to success. What is evident is that communication is a skill that will need to be practiced over time. One can only become better at it the more time that is invested in it. Being able to listen and speak supportively and concisely will not only make you a better leader, but a more empathetic person. Think of former President Barack Obama—his ability to establish an empathetic connection with his audience while maintaining precise clarity is what makes him an exceptional speaker. Being able to understand and share the feelings of another will allow you to make decisions based on compassion as opposed to strategy or monetary gain, thereby humanizing your entrepreneurial pursuits. Above all, clear communication is the root of all successful relationships and high performance. Truthful, authentic communication creates community, deeper understanding, respect, and mutual values. This

24 Henrik Gustafsson et al, "A Qualitative Analysis of Burnout in Elite Swedish Athletes," *Psychology of Sport and Exercise*, 2008, vol. 9, no. 6, pp. 800–816, doi:10.1016/j.psychsport.2007.11.004.

25 Agnès Parent-Thirion and Christine Aumayr-Pintar, "1758 6th European Working Conditions Survey: Job Quality in Europe," *Eurofound*, September, 2018, doi:10.1136/oemed-2018-icohabstracts.36.

transparency allows for an environment of trust and loyalty, creating space for the development, progress, and fulfillment of one's purpose.

KEEP IT SIMPLE

The best way to guarantee effective communication is to place great value on simplicity. Concise, to-the-point communication no matter who you are speaking to or what the subject matter may be is an extremely valuable skill. This lesson has been particularly important in my relationship with my daughters. As you might be able to imagine, I have an occasional tendency to be quite complex when speaking. But I have learned that when communicating with my children, I get the best results when I simplify things.

Scientifically, this concept is known as Cognitive Load Theory (CLT), developed by John Sweller, which refers to the presentation of information at a pace and level of complexity the learner can fully understand. "Cognitive load" relates to the amount of information working memory can hold at one time. Sweller stated that working memory has a limited capacity and because of this, instructional methods should avoid overloading it with additional activities that do not directly contribute to learning[26]. In other words, just keep it simple.

Simple communication has helped me discuss complicated topics with my kids. When my wife and I wanted to start teaching them about the value of money and investing, I needed to find something they could

---

26 Paul Chandler and John Sweller, "Cognitive Load Theory and the Format of Instruction," *Cognition and Instruction*, 1991, vol. 8, no. 4, pp. 293–332, doi:10.1207/s1532690xci0804_2.

relate to and understand. I asked them to write down the entertainment and technology they liked engaging with. Their list included Amazon, Disney, and Apple, so we showed them how the value of each company changes in the stock market and used that to teach them about how business works and how their products can increase a company's value. We talked to them about how Marvel, another favourite, is owned by Disney and what that means from a business standpoint. Of course, it will take time for my elementary-age kids to understand concepts like what a shareholder is and the value of a share, but because we started this very simple conversation about it, we are guaranteeing their understanding of it in the future.

My own parents introduced these concepts to me when I was a teen. I had an investment account in trust with my mom that I had trading authority over. The experience taught me a great deal about saving money, the principles of compounding interest, and the types of tools I would need for my own growth, both with my own finances and as an entrepreneur. If we fail to introduce these concepts to our kids, they will never learn to appreciate them. The best way to do so is through easy, straightforward methods.

Keeping it simple helps to keep stress levels down and enhance performance. When you are in flight-or-fight situations, simplicity encourages you to remain calm. In order to simplify things to keep moving forward in such scenarios, it helps to have a very detailed, actionable plan you can refer to that clearly reflects your goals and the steps you need to take to meet them. Being prepared with things like checklists and clear deadlines allows you to count on your own performance to get through those types of situations, which will ultimately help you cultivate greater resilience because you are dealing with stress in a systematic matter.

GRADE-SIX LEVEL

I want to start with the basics. To be candid: I have a love-hate relationship with my uncle's go-to phrase.

My uncle, whom I consider a valued role model and mentor, is also my business partner. We spend a lot of time working together on complex, intricate matters and there are times when he asks me to conduct research. Naturally, I come up with pages and pages of very detailed and complex information. But no one has time to sit and read research papers or studies. What he needed—what we all need— is an abstract or summary. In those moments, he will turn to me and say, "Explain it to me at a grade-six level."

While it is oversimplified for my liking, I also dislike the implication that sixth graders cannot be exceptionally smart, mind you, at the time of this writing, I am currently the father of a daughter about to enter the sixth grade. But even though I dislike the expression, I understand the need for it because, sometimes, you just need to use the most straightforward language possible.

Since my childhood, I have not been able to solve the Rubik's Cube. Every time I tried to sit down and learn the steps and algorithmic solutions to solve it, there has been nothing but frustration. My oldest daughter recently learned to solve the Rubik's Cube with our brain coach, Justin Noppé, out of her own interest. Over a period of three weeks, she steadily learned the different steps, practicing each of the segments before advancing and within four weeks, she was able to solve a cube within ninety seconds, successfully beating Justin's record. She surprised even herself and, suddenly, I, too, was intrigued by watching her hand-eye coordination quickly manipulate one of the most challenging toys of my childhood. One afternoon, I witnessed her teaching my brother in the same incremental steps. It was impressive to listen to her uncomplicated explanation. Over the next week, Maheep became determined to prove to his eleven-year-old niece that he too could be "a cuber" and succeeded in solving the cube. This is just one example of how children have a brilliant ability to keep things simple. The older you get, the more complicated everything seems to become.

In the business world, people often feel like they need to use a lot of professional jargon in order to convey intelligence or expertise. International writing expert Helen Sword refers to this phenomenon as "Zombie Nouns," which are nouns made from other parts of speech known as nominalizations. For example, when you add a suffix such as -ity, -tion or -ism to adjectives, verbs or nouns, you have actually created a new noun that may have a very different meaning than what was intended, making things more confusing than clear.[27] In other words, it is often far less important to impress people than it is to engage in effective communication, and that is where the importance of simplicity comes in. No matter how technical a field is, unless everyone involved understands the problem at the most basic level, nothing is ever going to materialize. In this regard, explicit communication is critical.

Keeping things simple is also key to helping you understand, express, and advance your purpose because if you cannot convey your intentions and desired outcomes to others, you cannot expect anyone to be fully on board with your vision. You have to be able to articulate what you want to do and why you want to do it.

27 Helen Sword, "Zombie Nouns," *The New York Times*, July 23, 2012, https://opinionator.blogs.nytimes.com/2012/07/23/zombie-nouns/?mtrref=www.google.com&gwh=37CC7B4257FD5DD6DA753A65D041E5DA&gwt=pay&assetType=PAYWALL.

Before we dig into this, I want to be clear: I am not saying anyone needs to dumb themselves down. You can and should challenge yourself to step outside of your comfort zone. People should continue to think complex—the key is understanding the difference between thinking complex and speaking complex.

LEVEL THE PLAYING FIELD

Beginning with the first emotion your face displayed when you were a baby to how you responded to yourself while looking in the mirror after you woke up this morning, you have been training to communicate. Communication is the basis of every relationship and every action or interaction grows from this. Knowing this, why is it that communication, or lack thereof, is often a source of conflict and the cause of undesirable results among individuals and within organizations?

Successful communication begins in any environment with the leadership establishing common cultural rules around how people are expected to speak with one another. This is similar to athletics, where a team may begin the season by assessing everyone's skill levels to reestablish a baseline or recalibrate positions and plays for the season. When people feel as if they have to impress everybody all the time, it is usually because the company has established norms that require them to converse in a certain way. In these environments, people are not encouraged to admit when they do not know something. They are expected to always have the answers. The risk you run in such an environment is creating a pool of people operating under the "fake it 'til you make it" theory. This would be very difficult to do on the sports field and would only highlight where the weaknesses are among the team. When people are not being truthful, no one knows who they can and cannot trust and the entire environment becomes toxic. As a result, tensions rise and negativity sets in, causing withdrawal and disengagement. Basically, faking anything is a great way to set both yourself, your team, and your company up for failure.

At my company, asking questions was not only okay, but also encouraged. We had multiple employees who came from different industries and not everybody shared the same universal language around biotech and life sciences. When people would employ terms commonly used in drug development like "ORR" or "PFS" or "Phase 2," not everybody understood, leaving some people embarrassed or pretending like they knew what everyone was talking about. To get everyone on the same page, we put together a glossary of key terms in our company handbook for new hires to refer to anytime they needed it. We did not encourage jargon, but we knew those words and phrases would inevitably pop up in conversations and we wanted everyone to feel included and prepared.

We also expected people to ask clarifying questions. In a meeting, if someone did not understand something that was being said, they had unspoken permission to interrupt and ask for a deeper explanation. It was casual—"Hey Mary, can I ask you a clarifying question on what you were just talking about?"—and everyone knew it was something that could happen at any time. We were firm believers that no question was a dumb question. Throughout the company, people knew degrading anyone for asking a question was simply not tolerated and, therefore, directness was expected rather than viewed as a threat. Even if someone inadvertently caught someone by surprise with a question, there was not a sense of being put on the spot. So, while it might drive me crazy when my uncle says he needs to understand something at a grade-six level, I also appreciate it because it allows me to ensure I fully understand a concept and can teach it to somebody else. If there is anything I missed, I know he will hold me accountable and ask for clarification to ensure that we are both on the same page. This kind of communication creates an environment of trust and loyalty, which can only lead to workplace fulfillment. It breeds positivity and during challenging times, having these type structures and shared understanding and meaning will prevent communication breakdowns.

My company was so particular about communication, we even addressed how we spoke to each other in emails. Whereas some people liked to pepper their emails with pleasantries before getting to the reason for the message, others preferred to skip niceties and get straight to their point. We found when a member of the former group received an email from a member of the latter, they often were hurt. They assumed the sender was mad at them, which I can appreciate. Imagine you are a new employee who recently joined a company and received a one-line, no courtesies note from the boss—your guard would most definitely go up. We certainly did not want people getting bent out of shape about things like that, so we made it clear across the company that if someone sends you a quick note with no greeting or sign-off, there is no negative intention behind it. We work in a fast-paced environment where information needs to move quickly. We were all working for the same outcome and brief email communication was in no way indicative of a lack of appreciation or respect.

REPEATING YOURSELF

Effective communication even plays a role in how our bodies self-regulate. The central governor model of exercise regulation is a process by which the brain regulates exercise performance by continually "talking" to different parts of your body. While you are working out, your brain is communicating with your muscles, your respiratory system, and your circulatory system, all in an attempt to control the way you are pushing your body.

BREAKAWAY

REIMAGINE TRAINING COMMUNICATIONS

"The highest compliment that you can pay me is to say that I work hard every day, that I never dog it."
—Wayne Gretzky

Work capacity is the amount of training you can perform, recover from, and adapt positively to after each session. High work capacity means you can train harder and for longer periods of time, avoiding exhaustion and overtraining.

I subscribe to the laws of doing more so your body can learn to adapt to a higher work capacity. By mixing up routines and getting out of your comfort zone, you are challenging your muscle memory and neuromuscular communication to send fresh signals, forcing the body to adapt to new conditions. People seem surprised when they plateau doing the same training routine, with the same repetition and set scheme they have been doing for several months, if not years. I am sure you can relate to the 3x12 strength training scheme you did when you were in high school and still repeat to this day, except you have not made any exponential gains or, depending on your frequency, may be hovering at almost the same strength. I can relate to this because twenty-five years ago, one of my swimming coaches introduced me to an ab routine that was ten different core exercises, twenty reps each, totaling 200 reps. To this day, I occasionally whip out this routine as a finisher to my workouts, but it has become so easy. Meanwhile, I am nowhere near those washboard abs I had at fifteen years old. When this happens, we have to work harder, longer, and mix up the training. Why? Because it increases the amount of training, stress, stimuli, and progressive overload your body can tolerate and positively recover from at the end of the session, so you remain in the Adaptation Phase.

In short:

1. New communication networks in the body are necessary to see improvement.

2. Greater work capacity means you can handle more training stress, which means you improve faster.

Lowered work capacity means you can only handle a lower amount of training stress, which means you improve, but at a slower rate. Increasing your work capacity is the only way to continually improve. Athletes can only ever claim they have reached the "genetic ceiling" of their physical ability when they no longer have the ability to increase their work capacity.

The work capacity sink analogy has been referenced by several different strength and athletic experts. Imagine the water coming out of a tap of the sink as your training stress. Now imagine the amount of water released into the drain is your work capacity. When training with a low level of stress, such as a steady 5K run or a manageable thirty-minute weights routine, there is a slight trickle of water coming out of the tap (training stress) and it does not take a large drain hole to drain the water (work capacity and recovery). As you become stronger, quicker and have more endurance, you will need more training stress to improve. That 5K run and basic strength training routine is no longer enough. You need more water to come out of the tap, but equally you need a wider hole/pipe to drain it.

A small drain means you can only tolerate a small amount of water. The wider your drain, the easier the water flows, the more water you can put in your sink, and the greater the improvements you will see, all the while ensuring your sink does not overflow (overtraining).

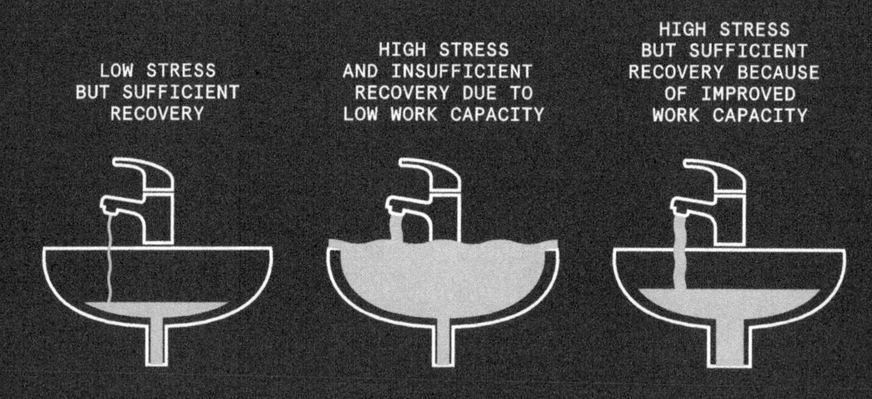

Work-Capacity Sink Analogy

It is paying attention to fatigue, intensity, and duration, factoring it all in when communicating with the different parts of your body. The more you repeat an exercise, the more your body learns and the better it is able to prepare for the next movement. When you are running a 10K race, this is the reason why you are able to go all out at the end. At this point in a race, you would think communication is breaking down between all your biological signals because this is probably the most physiologically challenging moment in the race from an exertion standpoint. You are processing many mixed signals and the muscles are fatiguing due to the lactic acid buildup, yet the brain is wanting to power through to the end of the race. You would assume your body would be exhausted by the time it hit that tenth kilometre and it might be, but by managing its energy levels and conservation based on the training, it is able to hit peak performance right at the end. In this kind of environment, there is absolutely an opportunity for negativity to set in, causing the body to withdraw and disengage from top speed. However, with proper repetition, the management of expectations, and the central governor model of exercise regulation theory kicking in, you can finish strong. In athletics, the more you teach the body to know what to expect when you exercise, the more likely you will be able to rely on that ability to perform in that last kilometre.

The same idea applies to communication in a business environment. In challenging times, communication breaks down and is often replaced with criticisms, complaints, and disapproval between individuals. This kind of environment also causes withdrawal and disengagement. The more you work on developing expectations around clear communication, the more likely you will be to elevate yourself and your team during the challenging times, counting on them when you want to put the pedal to the metal. A lack of candour impacts an individual's or an organization's ability to perform optimally. When individuals are proactively supported in telling the truth without fear of repercussions, problems can be identified and addressed early on and leaders have the information they need to make good decisions. Alongside honesty being the best policy, the simpler everyone communicates, the more effective they will be in reaching the outcome they are looking for, building resilience among individuals as well as within organizations.

As a leader, it is important to remember that for some people, engaging in a new way of communicating will come naturally; for others, it will take a great deal of repetition before new habits form. Introduce shared vocabulary, communication governance, and radical candour as tools meant to reinforce clear communication, structure, and trust. As the central governor theory dictates, the brain will learn what to expect as the territory becomes more familiar, guaranteeing the ability to perform when it is required.

INAUTHENTICITY

Perhaps the most important element of effective communication is honesty. A commitment to being overwhelmingly honest is the foundation to accountability and resilience. When you are not forthcoming about your purpose, intentions or capabilities, you set yourself, your team, and your entire company up for failure.

Before we get into a discussion about the deficiencies of inauthenticity, there is a minor caveat and small merit to faking it: the belief that you can accomplish a task without having undertaken it or that you have the ability but perhaps not the experience to hold a certain position can give you the confidence boost that you need and, therefore, your foot in the door. This is not necessarily a bad thing. We all have to start somewhere, and we were all beginners once. But this is as far as faking it can be beneficial—it can get you an opportunity, but you still need the skill set and the abilities to make things happen. In this sense, faking it can get you only so far, but, sometimes, that is all you need to prove to people that you have the capabilities to put your money where your mouth is.

However, if you fail to produce results and are grasping at straws to keep the charade going, you are purely operating in the dark. You are giving yourself a very low chance of accomplishing your goals while simultaneously guaranteeing damage to your integrity. Why would anyone voluntarily do that? You cannot make a decision that has any kind of impact on others if you are not confident in your own ability to make it. You cannot make promises without knowing you can keep them. You cannot rely on sheer luck, which is essentially what you are doing when you are not genuine about what you can deliver. Otherwise, you are someone who expects to win a 10K race without ever having run a day in their life and the only way you are going to be the first across that finish line is by cheating.

Faking it is linked to the value of transparency. The more you withhold information, the more you become trapped by it, losing time and energy trying to conceal things. This world is rich with opportunities for people to avoid confronting problematic work-related issues, ultimately contributing to an overall loss of morale and erosion of trust. Do not be that person. I know and believe that as a world-class Corporate Athlete, you are better than that.

TELLING HARD TRUTHS

Leaders do not always face easy or pleasant choices. Sometimes, they try to avoid telling the truth until it is too late. How do we eliminate the urge to stay silent and instead be brave enough to create an environment for telling the hard truth? While it might be tempting to take the easy way out, in the end, the choice to have courageous conversations will garner respect. People respect fearless, decisive leaders who are not shy about having difficult conversations. The more you demonstrate this skill, the more likely the people who work with and for you will follow your lead. The more you avoid the truth, the greater the price you are likely to pay in terms of loss of loyalty, frustration, and distrust.

History is not kind to dishonest people. Remember Bernie Madoff, the crooked financier who went to jail after swindling clients of billions of dollars through a Ponzi scheme? Imagine the pain he could have avoided had he applied the same amount of energy toward a legitimate business. The actions of liars almost always have a domino effect on others. Just look at Lance Armstrong—his doping scandal not only impacted every athlete he competed against, but also caused negative fallout with The Livestrong Foundation, the nonprofit he founded dedicated to supporting people affected by cancer. In the aftermath of his actions, he tried to separate himself from the organization, but it had already been caught in the crossfire and lost a great deal of revenue. Lance's story presents an interesting moral dilemma. He brought so much to the sport of cycling and to cancer awareness and philanthropy. On the other hand, he has been linked to such extreme negativity with the sport. Yet, he did confess and recognize the pain his actions caused and has been trying to rebuild trust with the people who believed in him. Lance's story shows that if you are living without authenticity, you can change course. You can stop your negative behaviours, confess to your indiscretions, and embrace who you are truly meant to be. He is not out of the woods yet and I am sure many people still take great issue with what he did. But I do not believe his poor decisions negate all the good he has been able to do in his life, particularly in regard to his commitment to advancing cancer research and speaking up about his mistakes.

In 2012, one year into starting OncoSec, I was on the brink of insolvency. We were down to fumes in our cash position but on the cusp of starting three clinical trials. As the CEO, there was an expectation of not losing any momentum, delivering on our important milestones for our new experimental skin cancer immunotherapy, and keeping the team motivated and focused on initiating several clinical trials. The reality of being an early-stage biotech company developing a breakthrough therapy for cancer that was at least a decade away from commercialization was stressful, even more so when it became clear in our reduced balance sheet that our finances would not be enough to finish what we started. There was a lot of pressure. I had the dual focus of outwardly keeping up the valuation and working on a capital raise and inwardly making some necessary cuts to stretch the remaining cash as far we could, which involved convincing my team to take the risk of deferring salaries and benefits until financing was completed. In practice, I find that challenging circumstances are also an opportunity for positive transformations.

In order to manage the external goals, I invited my scientific cofounder, Dr. Adil Daud, to join a critical investor meeting with Ayer Capital, a prominent biotech venture capital firm led by physician-turned-successful biotech investor Dr. Jay Venkatesan. Dr. Daud was able to succinctly explain the science and differentiation of the OncoSec technology and its wide application across solid tumours and I was able to focus on how our team was going to have the operational discipline to execute our ambitious clinical development plan for three simultaneous Phase 2 clinical trials. That meeting and the subsequent due diligence led to Ayer Capital taking a lead role in financing that was convincing enough for three other prominent biotech intuitions to invest alongside—if Dr. Venkatesan did the homework, it was a positive sign for others to proceed. On the internal front, I had two close team members, my CFO and the controller, instill confidence in the plan. In the end, we closed a day before posting a negative balance in our bank account. After that roller coaster experience, I had a better appreciation for the importance of my role as a leader and what the different stakeholders expected of me, the investors, my staff, and even Dr. Daud, who was counting on his research being available to patients. This was a defining, resilience-building moment that called on me to keep a sense of calm to convince the stakeholders and the team around me to believe in our mission.

By harnessing authentic communication and the use of persuasion, I was able to turn the ship around. If I did not effectively communicate with my CFO and controller to be comfortable with the risks, the alternative would have been to lay off the entire staff until things stabilized. With both these individuals supporting me, we were able to lead with conviction, not only by the collective leadership reinforcing the upcoming changes, but also by our decision-making and role modelling, as we encouraged our team to stay committed to our cause. We know from research that human beings strive for congruence between their beliefs and their actions and experience dissonance when these are misaligned. Believing in the "why" behind a change can therefore inspire people to change their behaviour. In practice, however, we find that many transformation leaders falsely assume the "why" is clear to the broader organization and, consequently, fail to spend enough time communicating the rationale behind any changes or efforts.

When you are in a leadership position, it is expected that you will be judged and criticized. If you are not, you are probably not doing your job. When you accept this role, you voluntarily agree to be under the microscope. Because every decision you make affects others, there will always be someone who will have an opinion about your decision. As a CEO, I understand what it means to have a thick skin—I have to be able to take a hard line on issues and know where I will not compromise. In order to achieve the buy-in and trust during those times, maintaining integrity has been the key. Anyone who is a leader, however simple and onerous that role sounds, must identify ways they can encourage more candid feedback in order to cultivate greater resilience.

RADICAL TRANSPARENCY

Billionaire hedge fund manager and philanthropist Ray Dalio is a proponent of what he calls "radical transparency," or complete openness. The concept is based on the idea that in order to have successful communication, a company must be able to share *everything*, especially the things that are the hardest to share. In order for it to work, leaders must ensure those who are given radical transparency recognize their responsibilities and weigh things intelligently. Those who are not able to handle it well are either denied the right or removed from the organization. Obviously, they should not share everything with everyone, as sharing sensitive information with the organization's competition would be counterproductive, for example. But when radical transparency is handled correctly, it results in meaningful relationships and meaningful work that are mutually reinforced, especially when supported by truth and transparency.

Allowing yourself to be completely open and expecting the same from others might seem daunting at first but the end results can be inspiring. "People spend too much time calculating the risks that come with being honest and too little time thinking about the rewards," says Brad Blanton, author and founder of the Radical Honesty Network. Blanton suggests that "radical honesty is addictive. Once people discover the truth, they fall in love with it."[28]

---

28 Brad Blanton, *Practicing Radical Honesty: How to Complete the Past, Live in the Present, and Build a Future with a Little Help from Your Friends* (United States: Sparrowhawk Publications, 2000).

Radical honesty can also result in some of the most productive, energizing conversations among your team. This may sound contradictory, but constructive conflict in the boardroom is a rich, complex experience that stimulates creative thinking and challenges us to grow. I recently witnessed this during a board meeting for a joint venture that has equal representation from both joint venture partners. Constructively speaking, the positive behaviours in the room supported expressing perspectives, creating solutions, revealing emotions, and actively reaching out to the other party consistently. On the other hand, I could not help but notice destructive tendencies during negotiations, where one party was very interested in winning at all costs, displaying anger and retaliatory behaviour. In the end, we were able to come to a win-win agreement and resolution, but the process that led us to the final outcome was very frustrating. Before coming to the final resolution, I observed that our CEO, who represents one-half of the JV, used constructive conflict for the board and management to develop new ideas and realign the entire organization on our objectives. By encouraging courageous truth telling over a false sense of harmony, he invited everyone to engage more deeply in what is most important and what matters most. He quickly addressed the underlying problems and issues in the organization and helped to establish the team identity. By seeking a true way to solve a problem, face a challenge or overcome an obstacle, this approach almost always enlivens, clarifies, and empowers people to accomplish more. Instead of dealing with disengagement, dissatisfaction, frustration, and resentment head-on, do you tend to skirt the issue, avoid the decision, and press on, or do you choose truth and transparency to build a resilient team?

Building an environment where you can constructively tell people why their work is not meeting expectations or what they need to improve on takes time. As coaching and mentoring become ingrained, a culture is created in which performance improves, engagement deepens, and resilience strengthens.

Take, for example, how difficult it can be to encourage an employee to leave to find their passion elsewhere and then assist that employee in launching an endeavour that will lead to greater satisfaction and fulfillment, both personal and professional. I experienced this firsthand with a former employee. When I first met him, I was mesmerized by his intelligence and I wanted him on my team. I was able to woo him away from another company and was beyond excited to see what he could do for ours. But not long after bringing him on board, I soon realized we both had made a mistake. He was a brilliant person who, it became clear, was best suited to share his wisdom with others. His purpose simply was not based in corporate leadership. When we inevitably parted ways, he confessed that he always liked the academic side of things best and that is where he works now. To his extreme credit, he was able to recognize where he was able to add the most value and he is now doing the work he was meant to do. Resolving the situation took several difficult conversations on both our parts but, ultimately, those talks ended up benefiting us all.

At the end of the day, honesty is a learned skill. It is uncomfortable, exasperating, and at times, frightening. I encourage you to sit in the discomfort. I want you to practice this in order to become better at it, become familiar with it, and to make friends with it because this feeling will guide you away from dishonesty as a backdoor shortcut. I strongly encourage you not to be naive about telling hard truths. If you commit to doing the difficult work now, you are setting yourself up for great success in the future. With constant repetition, telling the truth will come so much easier and feel much less like tough love.

LEARNING TO LISTEN

Entrepreneurs love to be the smartest person in the room. We relish having all the answers and pride ourselves on our ability to articulate every detail related to our company. But as we work to build our career and establish our organization, we get so used to the sound of our own voice that any other can seem like a threat. Outside input makes us feel judged and defensive, no matter who the source or what the intent. We are working so hard to impress others with our own abilities and skill set that we fear that a louder, more informed voice might drown ours out. This mindset is something we all have to push past because it hinders our progress—achieving excellence in executing purpose requires successful collaboration. That means we have to learn to listen.

Listening promotes new curiosities. When you open yourself up to other perspectives and recognize their value, you enhance your own purpose. Unfortunately, the business world expects people to act like they know exactly what needs to be done and how to do it all day, every day, in every situation. For a lot of executives, this ends up being their Achilles heel. In these environments, having all the answers trumps authenticity and, ultimately, no one gets what they want.

I firmly believe being the smartest person in the room is the dumbest thing you can do. There are endless opportunities for us to learn from others in just about any interaction if we just listen. As a leader, you must seek out and create those opportunities whenever possible.

BE A LISTENING LEADER

The first step in creating a culture that values genuine listening is to allow people to not have all the answers. For example, if I, as a manager, ask someone a question about their area of study and they do not immediately have a response, it is up to me how I react. If I am living a purpose-driven mindset, I will first appreciate that our level of communication is so strong that they felt comfortable admitting to not knowing something, and second, accept that they might need more time to produce an answer. If I behave this way, everyone within my organization will know that simply offering any answer just to have one or giving somebody what they want to hear are not acceptable practices. In fact, such practices are harmful for organizations because they lead to building a house of cards upon a foundation of assumptions and mistruths.

Many times, people fail to listen because they are too concerned with how they come off in the exchange. I have witnessed scenarios where a bullshitter is speaking to someone else and both parties are just stroking their collective egos. Neither person is tuned in to the other; neither person is listening to what the other is saying. Such superficial interactions accomplish nothing and can be particularly dangerous when the bullshitter is a person who does not have the required expertise but believes they do simply because of their position. Any input they offer ends up diminishing the effectiveness of the overall team because we may end up making significant decisions based on nonfactual information.

As a leader, you have to work to avoid such situations, which requires recognizing that every person is wired differently. When our team took the Emergenetics assessments, it helped us realize how each of us communicates and listens. My results showed that I am very social and rank high for assertiveness and expressiveness. Other people ranked far lower in terms of assertiveness or expressiveness, and that is okay. In fact, it is good because now, I can use that data in meetings or other decision-making settings to get the most of my team in order to produce the best outcomes.

I might encourage a nonassertive person to speak up, knowing they might not otherwise. I might recognize someone's analytical ability and know that they are taking all the information in rather than seeming disengaged from what is happening. I might recognize that someone is more likely to want to hog the microphone, so I will pay attention and make sure everyone who wants to contribute has a chance to do so. There is also an Emergenetics app, so when I go into a meeting, I simply select all the names of the parties involved and instantly know the room's composition. If I see the group is 55 percent social, I will be more aware of this going into the meeting to ensure we remain on task and that the group is balanced with analytical and structural contributors as well. Now I have created an environment ideal for effective decision-making. An effective leader must listen to everyone while recognizing the strengths and weaknesses of their team in order to get the whole picture of what they might require in order to have a purpose-driven organization.

We also encouraged everybody to put their Emergenetics profiles on their office windows, so when anyone wants to speak to someone else, they could look at their profile and instantly know the best way to communicate with them. Something as simple as a visual representation of everyone's strengths and weaknesses can lead to more mindful, productive conversations.

LISTEN UP

The benefits of actively listening to one another are immeasurable in all areas of life but especially in the workplace. Placing value on listening fosters a culture of collaboration. It helps everyone become more comfortable with feedback. It gives everyone the opportunity to learn from one another. It also creates clarity, so you can avoid potentially derailing abstractions.

BREAKAWAY

TRAINING YOUR MIND

Whether or not you face a challenging workout or upcoming competition, your ability to remain calm and focused under any circumstance builds resilience. In stressful situations, we have a tendency to have a high level of alertness and irritability, but by nurturing strong and stable attention, we are better able to harness calm, focus, and clarity. The way to train your attention is through a practice known as mindfulness meditation. According to Dr. Jon Kabat-Zinn, professor of medicine emeritus and creator of the Stress Reduction Clinic and the Center for Mindfulness in Medicine, Healthcare, and Society at the University of Massachusetts Medical School, mindfulness is "paying attention in a particular way on purpose, in the present moment, and nonjudgmentally."[29]

Research suggests that by meditating regularly, the brain is reoriented from a stressful fight-or-flight mode to one of acceptance, a shift that increases contentment.[30] Research from Carnegie Mellon University has become the first body of work to demonstrate that even brief mindfulness meditation practice—just twenty-five minutes a day for three consecutive days—can mitigate psychological stressors.[31]

---

29 Jon Kabat-Zinn, *Wherever You Go, There You Are: Mindfulness Meditation in Everyday Life* (New York: Hachette Books, 2005).

30 Antoine Lutz et al, "Long-term Meditators Self-Induce High-Amplitude Synchrony During Mental Practice," Proceeding of the National Academy of Sciences, 2004, 101, 16369-16373.

31 David J. Creswell et al, "Brief Mindfulness Meditation Training Alters Psychological and Neuroendocrine Responses to Social Evaluative Stress," *Psychoneuroendocrinology*, 2014, vol. 44, pp. 1-12, doi:10.1016/j.psyneuen.2014.02.007.

Published in the journal *Psychoneuroendocrinology*, the study investigates how mindfulness meditation affects people's ability to be resilient under stress. It can be challenging to sustain mindfulness in times of stress or trauma, but when you use mindfulness meditation, you are able to stay calm, with a greater sense of tranquillity and serenity. This leads to increased optimism and happiness, which builds resilience.

Just as you are training your body to increase work capacity by strength, speed, and mobility, meditation training enables your mind to gain greater flexibility and strength. The same principles apply: the more you practice, the stronger the mind becomes and the better you are able to increase your capacity in all levels of performance as a Corporate Athlete.

Mindfulness is also an important practice that can change your general attitude. Maybe you have reached a plateau and you may be questioning your abilities, or perhaps your internal flame is running low. Most often, the real barrier lies in changing your mindset from a fixed mindset to a growth mindset, believing that you can and will make progress instead of the opposite. It is easy to fall into this trap. After you have acquired a lot of experience and knowledge in a particular field or sport, doing things faster and with more proficiency and settling into a comfortable mindset, any shock and adaptation phases may be harder to come by. Becoming complacent is a short-term solution that leads to long-term problems. If you remain stuck at this languishing pace, you are not building resilience. It is actually less time-consuming to embrace a growth mindset, one that assumes you are going to build on your strengths and push your capacity. Of course, shifting to a growth mindset takes effort, but once the intention is there, the work becomes a little bit easier.

In running, it can be as simple as missing a critical day in training. Think of a cool, fall day, when the morning temperature has dropped, the wind is howling, the rain is rapping the window, and you are snuggled beneath your blankets. You could sleep in, but in the back of your mind, there are so many mixed thoughts: This is a run you feel you need to do, but this may also be a good day to take off. Perhaps you are training or maybe you have not run for a few days and you are worried you will begin to lose your hard-earned conditioning. But knowing the importance of the run is not enough to get you out of bed today. You need to change your attitude.

The first step toward overcoming your motivational problem involves a heavy dose of introspection. Assume you are like me and find your resolve to run quickly fading when the weather turns nasty. Try to list the reasons you have for not wanting to run. The list might look like this:

1. It is cold.
2. It is wet and dark.
3. I have no idea where my waterproof runners are, so I think I will stay in bed until the weather clears up.

The process of bringing to the forefront each thought and emotion you have is an example of mindfulness. Being mindful is being attentive to what is going on in your mind. When you examine your list more closely, you may notice the first two items are different from the third item. Being cold, wet, and dark are not thoughts or emotions. The outside weather and light conditions are not things in your mind at all. They are things outside of your control, things that are part of this world we know and often find this time of year. Indeed, you knew of this possibility before you went to bed. On the other hand, the third item is listing the only thing in your mind: you can *think* about where you left your waterproof runners and remember where to find them. I hope this distinction between things in your mind and things outside it is intuitive. Being mindful means paying attention to your thoughts and emotions and it is through mindfulness that we can overcome the thoughts and emotions that prevent us from doing the things we need to do and building resilience in the process.

The Training Your Mind *Breakaway* provides some additional correlation of active listening to mindful meditation. When you bring active listening and attention to your work, you are more present and focused with people. You listen more attentively without your mind wandering and settling into judgement. Active listening is a very powerful way to build trust and create connectedness.

Listening lays the foundation for win-win interactions and there are several ways you can improve your own skills in this area:

ASK EXPANSIVE QUESTIONS

No question is a dumb question but few things annoy presenters more than obvious questions. When someone running a meeting gets nothing but the most surface-level questions, they know those in attendance either:

1. Did not prepare for the meeting,
2. Could not care less about the meeting
3. Were not actively listening to anything they said.

When asking a question, it is better to try and build on the ideas being presented rather than simply stating the obvious. Try to focus on asking what and how. Such questions prompt people to delve deeper into the information or reflect on a certain situation, making them feel more heard.

Few things can kill a conversation faster than a question requiring nothing more than a yes-or-no response. Never ask a question just for the sake of asking something because no one, particularly the person being asked, appreciates that.

SHIFT THE FOCUS

Everyone has found themselves in a conversation where they are focused far more on themselves than the person speaking. We spend the entire time getting ready to respond rather than actively listening. I have been there. In fact, I have sat in negotiations with a term sheet in front of me while the opposing side is making their argument and all I am doing is thinking about how I am going to respond. This is tempting because most people are so anxious about making sure their own points are made, they miss most of what the other people are saying. This type of half listening is fairly common and is the result of not being able to stay in the present moment. Everyone needs to be heard completely. Listening fully is an art and a really important skill to develop.

It is okay to take a second to process before you respond. We all need to become better at being comfortable with silence, as it has the power to communicate attentiveness and respect. Of course, this can be a challenge for people who love the sound of their own voice. People who like to hear themselves talk tend to dominate discussions and drown out others who are less vocal or who simply need more time to think before responding. Let things land. Share the platform. If your voice is the only consistent thing in a conversation, it is time to be more selective in terms of when and how you contribute to the dialogue.

It is also okay to finish a meeting without having all the answers. Table a topic if it requires new or different insight. Part of successful collaboration is listening to fresh perspectives and seeking out value in one another.

PRACTICE SELF-CHECKS

It is also important when engaging with others to critique yourself often. For example, I know I can occasionally have difficulty articulating my thoughts clearly. Therefore, I have to actively work on making sure I avoid being too abstract. I encourage you to also create a similar self-awareness around how people are receiving you and engaging with you. Think about your own strengths and weaknesses and how you can use or change them to become a better listener and communicator.

A good way to start is by paying attention to body language, both theirs and yours. Nonverbal cues can convey just as much or more than anything that comes out of your mouth. Avoid eye rolling, yawning, sighing or anything that makes you seem guarded or disinterested. Also pay attention to the body language of the person with whom you are speaking and adjust the way you engage with them accordingly. Dan Coyle, bestselling author of *The Culture Code: The Secrets of Highly Successful Groups* who studied different world-class teams such as the US Navy SEALs, refers to these behaviours as "belonging cues." They are a cluster of little behaviours that we do not normally pay much attention to, but they are things people do when they care about and respect one another. He writes, "belonging cues are behaviours that create safe connection in groups. They include, among others, proximity, eye contact, energy, mimicry, turn taking, attention, body language, vocal pitch, consistency of emphasis, and whether everyone talks to everyone else in the group."[32]

Whether you are at the conference room table or dinner table, everyone wants to feel like a valued member of a group and that their thoughts carry

32 Dan Coyle, *The Culture Code: The Secrets of Highly Successful Groups* (New York: Bantam Books, 2018): Chapter 1.

weight. Ensure everyone is getting a chance to speak and that people are paying attention to one another and making eye contact. It is imperative that body language is respectful and everyone feels heard. As a leader, it is your responsibility to encourage an environment in which everyone has the right to express themselves and no one has a fear of speaking up.

ADD A 'PLUS'

There is a concept in improv called "yes, and." When you are in a scene with a partner, you agree not only to commit to whatever idea they come up with, you also work to build upon it. So, if actor one says, "I cannot believe this weather!" actor two might raise the collar of her imaginary coat and say, "Well, we are in the North Pole! That polar bear over there gets it," and the scene is off and running. If actor two refuses or ignores the first actor's initial idea, the scene is dead before it even starts. In the workplace, I call this "adding a plus." It is a way of saying, "I embrace your idea, but what if we also did this in addition to that?" It is a way of listening aimed entirely at helping someone enhance their own idea.

I find this tactic particularly helpful when it comes to getting my kids to do what I want them to do, which is the never-ending plight of every parent. I take the time to listen to what they want to do, then I add a "plus." Recently, I wanted my daughter to join me in my workout. She was definitely not interested in my regular routine but instead of saying, "No, you are doing what I want to do. Deal with it," which would have resulted in her saying, "Forget it. I am not doing this," I asked her what she wanted to do and I really listened to her response. I decided to say yes to her recommendations and make my regular circuit the "plus." I created a workout that was a combination of what we both wanted to do, and we had a great experience. When we were done, she actually asked if she could work out with me again later that week.

To me, that conversation had so many dividends. It was time we spent together bonding. It showed my daughter my willingness to consider her ideas. It took a possible negative and turned it into a positive. This is just one example of how adding a plus to the conversation can open doors in just about any situation.

WALK THE TALK

In a world that is often full of constant, mindless chatter, refining your communications skills can only help your cause. With the multitude of electronic distractions that we deal with on a daily basis, the practice of dropping in, being mindfully quiet, and thinking before you speak will set you apart as a conscious, compassionate leader.

Life is always going to surprise you and, sometimes, you might just have to give a presentation in a foreign country with slides that are not in your native language. How do you recover? How do you ensure you will not miss a beat and work with the obstacles that crop up unexpectedly?

When you set yourself up for success by putting in the time to refine your skills, whatever turbulence you encounter will not be able to derail you from your goals or accomplishments. Every moment has led up to this one. How will you show up?

□

WORLD-CLASS CODA

- Assess your capacity to be an effective, persuasive communicator during transformative times.

- Is truth telling encouraged and rewarded in your environment or organization? Is a false sense of harmony being chosen over truth? Identify opportunities that will help you uncover hidden truths in order to build resilience.

- Remember that meaningful relationships and meaningful work are mutually reinforced, especially when supported by attentive listening, truth, and transparency.

CATAPULT FORWARD

- What courageous conversations have you been avoiding?

- What will be the payoff for having these conversations?

- What secrets are you harbouring? If this information was "leaked," what impact will it have on your relationships and your overall success?

Leadership is an action, not a state of being. Leading does not preclude you from setting aside your ego, asking questions or consulting others in order to reach your goals. Leading, in fact, is not about you at all—it is about understanding that you may not know everything there is to know, and this humility will ultimately help you. There will always be someone who is smarter, faster, more creative or innovative, and by recognizing this, you will be able to learn from these dynamic people rather than be stifled by your envy of them. Look to others. The answers you are looking for may be right in front of you.

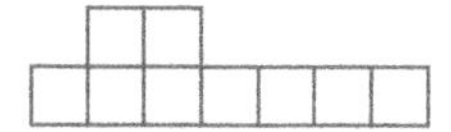

PRINCIPLE 8

RADICAL OPEN-MINDEDNESS

"You cannot transmit wisdom and insight to another person. The seed is already there. A good teacher touches the seed, allowing it to wake up, to sprout, and to grow."
—THICH NHAT HANH

Early in my biotech career, I was invited to sit in certain research and scientific meetings. It was a privilege but because I was relatively new to the field, I felt out of my depth, as I was learning on the job. During one particular incident, I was somewhat familiar with the operational and financial side of an HPV vaccine program the company was developing that was close to entering Phase 2 clinical trials but not as familiar with the science of the vaccine. As the head of finance and operations, my role was to follow up with the R&D team to discuss project updates and the budget forecast for the following year. At one point, the meeting deviated topics and, suddenly, it became extremely technical. Everything sounded very important. Since I was in charge of the meeting, I felt a responsibility to put something into action on this particular issue but I was completely floundering on the science. At this point, I had two decisions in front of me:

1. To address this concern in a structured manner to get the team back to the task at hand

2. To ensure that even though I did not understand the matter completely, to appoint someone who did so I could follow up with this critical concern

I summoned the highest-ranking scientist to summarize the key points for me and then delegated questions that allowed me to instigate actions linked to a timeline, asking the appropriate person to follow up. Within five minutes, we were all back on track. In this particular example, my meeting could have easily been derailed because I was very close to treading beyond my technical skill set, yet I was able to remain calm and focused on what I needed to get done.

DISMANTLING THE EGO

Entrepreneurs like to be the smartest person in any room and have a tendency to want to do it all themselves. They love to be heard, which can become incredibly self-serving. They feel like they know what is best and are not open to having any outside help. They do, after all, have the most comprehensive understanding of their company. They live and breathe their purpose, so why would they need anybody else to help support it? They do not need experts because they simply *are* the experts.

When self-serving leaders are experiencing success, they look in the mirror, beat their chests, and tell themselves how good they are. When things do not go well, they look outward and blame everyone else.

While incredibly common, this thinking is also extremely flawed. Leadership without an ego is a rare commodity. When leaders get caught up in false expertise, pride or self-doubt, it erodes their effectiveness. Ego gives people a distorted image of their own importance. It is understandable that leaders can fall into this pattern because, in the early days, entrepreneurs do have to do everything themselves. When working on a shoestring budget, you might just be focusing on getting from one month to the next and have incredible ownership in the cause. That is to be expected, but it is also important to remember that this is just a phase. As you start to get more support and things start ramping up, you have to expand your mindset and allow others the opportunity to grow with you and establish that same level of ownership. Not to mention, there will be many instances when you can delegate in order to hit your objective faster.

When things go well, it is an opportunity to look past your desk and share the credit. When you lead with ego, you put your own agenda, safety, status, and gratification ahead of you. A good leader recognizes different experts will bring different parts of their vision to fruition—this requires humility and the desire to bring out the best in others. Jim Collins talks about this in his classic book *Good to Great*. Collins suggests there are two characteristics that describe great leaders: will and humility. Will is the determination to follow through on an organizational vision, mission or goal that is bigger than you are. Humility is the capacity to recognize that leadership is about serving others instead of being served.[33] In brief, when things go right, give credit to others, and when things go wrong, ask yourself, "What could I have done differently?"

By expanding our thinking to the concept of being radically open-minded, we are better equipped to make good, effective decisions as world-class Corporate Athletes. Our egos and blind spots often are our biggest barriers to success. The ego is our insatiable desire to be the best in everything we do and to have others recognize this capability as a marker for our own individual growth. Our blind spots are the result of our seeing things through our own very limited and subjective lens. Both barriers can prevent the Corporate Athlete from seeing things as they really are. To circumvent them, the immediate corrective action is the idea of radical open-mindedness, which has been primarily promoted by Ray Dalio. He notes that "most people hold on to bad opinions that could easily be rectified by going above themselves to objectively look down at their situations and weigh what they and others think about a decision."[34] Consulting others is key to an unbiased perspective. Every world-class Corporate Athlete needs to have the ability to effectively explore different points of views and possibilities without ego or blind spots getting in the way. This nurtures optimal decision-making because now they have all the facts in front of them.

Doing this well requires the Corporate Athlete to seek out brilliant people who are better and more efficient at getting specific things done. However,

33 Jim Collins, *Good to Great: Why Some Companies Make the Leap...And Others Don't* (New York: Harper Business, 2009): 102-105.

34 Ray Dalio, *Principles: Life and Work* (United States: Simon and Schuster, 2017).

they do not need to know everything just because of a specific job title. Rather, a Corporate Athlete's job is to bring out the best ideas from each person on any given project and assess what is the most important for the organization. Often, this means working with people who disagree with them in order to see things through their eyes in order to gain a deeper understanding. Disagreement can be a very good thing. Some leaders like to get multiple points of view and, oftentimes, that is when real innovation occurs. Granted, other leaders prioritize a culture of harmony, and while that can seem appealing, I can tell you from personal experience it is not necessarily ideal.

When I was a young CEO, my C-suite was made up of a team who had twenty-plus years of seniority and very strong personalities. I was not experienced enough yet to understand how tension or discord could be helpful. Eventually, we put in place communication principles that required radical candour, and I soon learned that when people have the freedom to speak openly, trust forms. It is the job of the leader to encourage this kind of communication and not shy away from it. Use your coaching skills and assess everyone's styles and strengths to bring out the best in everybody. Doing this will raise your probability of making good decisions and will also give you a meaningful experience in your career. If you can learn radical open-mindedness and practice thoughtful disagreement, you can exponentially increase your learning.

Jim Heppell, my first boss, is an example of someone who did this very well. When I was twenty, I was hired at his corporate finance law firm as a clerk. Jim held a weekly lunch at which everybody was encouraged to present any interesting legal problems as well as challenges they had encountered within their current caseload. Attendees also discussed recent updates to laws or any changes they were seeing in the industry. Those lunches became an important learning experience where highly specialized people had an opportunity to share their own expertise and learn from others. Different experts from all areas of law brought varying perspectives that enabled the sharing of information and the weighing of different options to determine the right path to follow. It was not a sign of weakness to gain insight from the other lawyers, even when their disciplines were so different. Everyone in that room, except me, was a partner who could do things alone, but it was obvious to them that going it alone could only take them so far. Once a week, there was a moment for everyone to let their guard down and triangulate to figure out the best course of action, which helped the law partners see any weaknesses objectively and to compensate for them. It established more trust between the partners of the firm. Collectively, their boutique corporate finance law practice continued to flourish.

Similarly, my close friend and former colleague Iacob Mathiesen, the former chief scientific officer at Inovio, is another prime example of an individual who is still able to rally the best team for the job. When we acquired his company from Norway, he had already built a large team of scientists, including some who moved with him to California, where our company was headquartered. By virtue of the experimental stages of testing ideas and hypotheses, R&D teams in biotech may seem unproductive relative to the operational discipline that comes along with the later stage of product development. A good R&D team is well-organized and Iacob did an exceptional job of building a group with specialized disciplines that allowed different individuals to work together. This added substantial value, as research ideas made their way into the development pipeline. Iacob was gifted at both delegating and helping people become specialized in particular areas. They completed projects with great efficiency and the quality of work was unparalleled as they tackled some of the most challenging diseases. Good leaders know how to construct teams based on the organizational needs of a company while building on and benefitting from the skills of highly specialized people. When this is done right, people are more than just cogs in the machine. Our ability to succeed when working with others who want the same things is much greater than our ability to get these things done by ourselves.

Like athletes, entrepreneurs need experts. Take adventure racing—without an orientation specialist, there is no organization, no structure, no schedule. You need someone who excels at kayaking, someone who excels at mountain biking, and someone who excels at hiking and trail running. It is exactly like the business world, where you need to rely on specialized skill sets that propel the collective team closer to its objective. But experts do not always have to be the ones doing the heavy lifting to get you to your goals because, sometimes, just their example will suffice. For example, Alexander Popov is considered the greatest sprint swimmer in history. Popov simply had a stroke style that was so specific, my fellow swim team members and I studied it closely in an attempt to emulate his efficiency to cut through the water. There was a logic or rationality to his stroke technique that made it very teachable to other swimmers. We studied how one of his arms entered the water at the exact moment his other arm was exiting. We studied how his high reach allowed him to catch the water at the top of his stroke. We designed drills around his technique. Even though he was not a part of our team, we saw him as a resource who could offer tools to help us become more adept.

No matter how it happens, collaboration with experts can strengthen your foundation, and finding the right experts to help advance your mission can elevate your work and purpose beyond even your own expectations. When you put your focus on others, you are bringing them under the tent, giving them accountability and buy-in to the objective, and helping them reach a shared goal. By providing them with this level of personal ownership, it allows for the sharing of responsibilities. You are strengthening your own resilience as you create a network of people who can help you navigate any hurdle. Because the power of a group is much greater than the power of an individual, you will undoubtedly become more proficient in your approach to running your businesses. There really is strength in numbers.

THE GIFT OF TIME

Many experts are extremely busy and do not have much time to offer you. But when you are lucky enough to have them give you their time, you know it can be of the utmost value. The late Dr. Holbrook Kohrt was the smartest man I have ever known. He unknowingly became a missing piece of a puzzle I had been voraciously trying to solve and because of his expertise, he has helped countless melanoma patients recover and return to their healthy, productive lives.

But first, a little technical background: the former standard of care for skin cancer—chemotherapy—was toxic to the body and also had a dismal response rate, around 11 percent. Bristol Myers launched an immunotherapy treatment in 2010 with a response rate around 20 percent. However, it came with its own issues because when you amplify the immune system, a host of negative side effects can occur. In this case, it was leading to the presentation of other types of skin cancer. So, in 2016, a next-generation immunotherapy came along called Anti-PD1, which blocks the body's checkpoint pathways that alert the immune system to cancerous cells. Blocking the pathway forces the immune system into overdrive so it can tackle the cancer, and Anti-PD1 brought the response rate up to 33 percent. But 67 percent of patients were still not responding to it. However, the drug we were working on at the time was able to convert those nonresponders to responders. Essentially, our drug helped to turn on the immune system, so when Anti-PD1 was used, the right immune cells were present and ready to kill the cancer cells. Our drug had a success rate of 50 percent for the nonresponders.

We were on the verge of a significant breakthrough if we could make the drug widely available. However, we risked becoming obsolete before we even had the chance to do so because Anti-PD1 was taking over as the best immunotherapy ever developed, as it was not only effective in melanoma treatment, but also in the treatment of almost all other cancers. The trade name of Merck's version of the drug is Keytruda and the Bristol Myers equivalent is known as Opdivo.

Dr. Robert Pierce, my company's chief science officer, had worked on Keytruda at Merck. He joined my company when we recruited him along with three other Merck executives. I knew our team had a lot of great insight and nobody else was focusing on nonresponders at that time. Everybody was trying to focus on making responders respond better and not converting nonresponders into responders. We knew we could not compete against Merck or Bristol when they already had the winning drug, so being a second-line therapy or even a primary therapy for nonresponding patients was one of our top goals.

The clinical trial was a last-chance attempt for our company. I needed someone who was untouchable in terms of their brilliance around designing such trials to guarantee success. There was no room for error and we had to act effectively and efficiently. Dr. Pierce had previously worked with a brilliant clinician named Dr. Holbrook Kohrt and suggested he might be the person I was looking for to perfect the trial. Dr. Pierce happily made an introduction. Little did I know that Dr. Kohrt was about to change my life.

Dr. Kohrt was born with a rare form of hemophilia and his experience with the disease ignited within him a passion for solving problems in the medical world. He had an unparalleled gift in terms of understanding how to design a clinical trial in order to extract the best possible results. He was living in London at the time, so I asked if I could visit him to talk about the trial. He had intimate knowledge of melanoma and an excellent grasp of the cancer immunotherapy landscape. I wanted to get assurance from him on our penultimate trial that could lead to an approval of a therapy where late-stage melanoma patients had no other option available to them.

In January 2016, we met at the Ham Yard Hotel in London and I explained the situation to him over breakfast. Dr. Kohrt, frail but alert, sat before me with a cup of fresh ginger juice warmed to his exact specification. I remember thinking, *Don't ever forget this moment*, because I knew I was incredibly lucky to have this meeting with him. He listened intently, then took out a scrap piece of paper. On it, he scribbled detailed instructions on how to design the trial in order to make it a win.

He absolutely nailed it. Our team had designed the project 80 percent of the way and Dr. Kohrt provided the critical 20 percent we needed to have confidence in the development plan and trial design, eventually leading to the approval of the product. He designed a clinical trial that he knew had the best chance to get us the desired results and the competitive differentiation the company needed to succeed. He was the expert we needed, and rather than waste time trying to do it all ourselves, we relied on his expertise and guidance when we needed it the most. We were able to move quickly in the decision-making to finalize the study design because the majority of important factors had already been considered. The valuable insight offered by Dr. Kohrt made the final decision-making very swift.

When you have access to experts who hold key information, there is often pressure to act quickly. This is also a consequence of time. The one important thing I have learned from this experience is this: When you have qualified experts in your corner, you should feel confident that they have given you the best information to proceed. This is the entire concept of turning to people who know more than you and trusting that they can elevate your cause. In the case of Dr. Kohrt, I had zero hesitation to implement and proceed using his knowledge. When things are time sensitive, the confidence you have in your experts should give you the green light you need in order to plough ahead.

We talked about a lot of other things during that breakfast, and at the end of the meeting, I asked Dr. Kohrt if I could do anything for him. He gave me a very subtle smile and said, "Make this worth the effort." Sadly, Dr. Kohrt passed away just a few weeks after my meeting with him. He too had limited time. An average person lives about 30,000 days, if they are lucky. Still, that number does not feel like a lot of time, and many of those days we probably will not remember. Even so, with the inevitability of death, we rarely think we are living on borrowed time. Dr. Kohrt understood the value of his time on this earth—he did not squander it, making the most of it up until the very end. If what you are doing is not fulfilling you, shift gears, make changes, and try to find the things that make your life more meaningful. Every day is an opportunity to take chances because later is not guaranteed. Make everything you do worth the effort.

The clinical trial is now his legacy. He remains an example of an expert who did not have much time to give but used the time he did have to make the most valuable contribution possible. He was a deeply compassionate person, equal parts genius, down to earth, and funny. Even though I met him for business purposes, he was charming and inspiring, and when he spoke, it was my privilege to listen. The time I had with him will always be something I cherish, and his impact on the world of medicine is everlasting.

THE POWER BEHIND DECISIONS

When you are trying to achieve your purpose and live a purpose-driven life, you must consider all the resources available to you. Within any business operation, it is critical that you listen to experts, coaches, mentors, and even yourself. It is important to establish an effective structure for decision-making that takes into account the merits of each person's ideas. Ray Dalio refers to this as the "idea meritocracy," explaining that in a proper decision-making system, the best ideas win.[35] It is simply not possible for everyone to be in consensus all the time, and a leader needs to take into account all views in an open-minded way. At the same time, the ideas need to be filtered by placing each view in the proper context of the experiences and track records of the people expressing them.

Corporate Athletes must rely on their own experience in order to effectively use an idea meritocracy. This is not meant to be contradictory to the principle about leading without ego. The difference is remembering that if you cannot successfully do something on your own, do not think you can tell others how it should be done. We have all witnessed the blame game by people who repeatedly failed at something but are adamant about sharing their opinion of how it should be done, even though their opinion is at odds with those who have repeatedly done it successfully.

Corporate Athletes should always value facts over opinions. Everyone has opinions and, unfortunately, when they are not based on facts, they are

35 Ibid.

often bad. There is a very clear demarcation between facts and conjecture. Even your own opinion can be very damaging and harmful. It is best to remain as objective as possible as a leader. In biotech, people rarely ever share their experiences, a trend that is instilled in the culture. This is unfortunate because we can learn so much from one another. The reality is data does not lie—we can challenge the data, or we can make note of it and choose to do things differently.

LOOKING BEYOND THE EXPERTS

As I noted earlier, the team who reported directly to me was composed of highly qualified people. They also had their own teams of high-performing people. These teams, in their unique pods of specialty, included the people best positioned to really make things happen—mid-level managers. The C-suite was the group responsible for the business as a whole, but the mid-level managers ensured that the goals, outcomes, and objectives were met.

Mid-level managers are very motivated. They are hungry, clear about their expectations, and capable of making major contributions. In fact, they are usually the ones doing the work that has the most impact on any company's success, though most people might not realize this. Mid-level managers see it all. They form the heartbeat of any organization. Their purpose-driven approach makes them ideal for getting results when they are needed the most and their roles and motivations are usually very closely aligned with experience and responsibilities. Cultivating relationships with such purpose-driven people can have impacts beyond your wildest imagination.

The C-suite are not necessarily on the front lines, but they are contributing to the day-to-day, behind-the-scenes progress of the company milestones.

For example, a clinician makes a schedule, then a nurse prepares all of the components that allow for a patient to be treated. Once that patient is treated, biopsies and other tests are conducted and data is aggregated. That data goes to the staff of the chief medical officer, who compiles the information to be analyzed and the chief medical officer may then follow up on specifics with the team or clinician for interpretation before it is finalized and presented to stakeholders. The C-suite are working fully with their teams, who are most connected to the physical frontline work.

Systems theory, the study of interdisciplinary functions as they relate to one another within a larger, more complex system, explores this concept as it pertains to how people in certain positions are responsible for helping others progress. It is an approach employed by many companies and organizations dedicated to helping ensure the success of their operation and employees. The most important decision as a CEO is hiring the right team and, similarly, the C-suite also has the ultimate responsibility for hiring the people best fit for the job. Each person at each level is accountable to someone and has someone they report to. Systems theory can also be applied more broadly in an organizational setting. When Roche acquired Genentech, they had two different teams—the "red team" in the US and the "blue team" in Switzerland—that were doing the same work in parallel. It was purposely done in this manner to create a certain degree of healthy competition and guarantee the best outcome. The internal competition was a development plan insurance policy for the company, where only the best succeeds and proceeds to the next stage of development and there is a backup in case a second option is needed. In these scenarios, the best product is the one that is left standing. The people involved are very aware of what the mission is and how they are forwarding the cause of the overall organization. For a company like Roche, this provided a macro picture of accountability. The same concept can be applied to athletics. Under systems theory, organizations share a commonality but still operate as individual teams, much like how professional athletic leagues are set up. The NHL, NBA, MLB, and NFL all belong to these superstructures but are also competing against one another.

Once goals are placed in the hands of the right person on the team and it is made clear that they are personally responsible for achieving those goals, there should be excellent results. If you are a Corporate Athlete or a manager, you have to be confident that you are up to the given task as well. Those who are responsible for the goals, outcomes, and processes at the highest levels need to choose their workers wisely and manage well. Senior managers must be capable of higher-level thinking and be able to lead and steer their team toward success.

COMING FULL CIRCLE

While I was running OncoSec, I had a broader appreciation for how teams on the ground really can mobilize the work. This was partly thanks to my relationship with several experts, such as Dr. Daud.

At the time, Merck's drug was regarded as the standard of care for patients with stage four melanoma, meaning if you were lucky enough to receive their drug as a cancer immunotherapy, you had a high chance of resolving your disease. But in about 40 percent of patients, the drug did not have a response. Such cases were doubly bad for the affected patients because in addition to the drug not working, their immune systems had also been compromised. Yet there had to be a solution—and we found one. When we combined our therapy with Merck's drug, it ended up converting the nonresponders into responders. This trial is still ongoing at the time of this writing, with more data expected in the coming year.

Around the same time that we were seeing such successful outcomes, I received some devastating news. My longtime best friend's mom, Mrs. Rosario, had been diagnosed with melanoma. This woman was like a second mother to me—her son and I were the closest of friends in elementary school. They lived right across the street from me growing up, and Mrs. Rosario would often watch me during the early-morning hours before school when my mom needed to get to work. She had always encouraged and supported me, so when I learned of her illness, I knew I had to do anything I could to help her.

The treatment we were offering was not yet available to her in Canada, but I refused to let that hurdle stop me. I started by talking to our friends at Merck. Through the course of many conversations with several people within the organization, including Dr. Daud, and working with her doctors every step of the way, we were able to find a way for Mrs. Rosario to get treatment through the BC Cancer Agency.

I am beyond happy to say that at the time of this writing, she is in remission. That would not be possible had I not listened to my team and believed in their abilities. The result of those interactions had a direct impact on someone's quality of life. We were able to help a person who fostered confidence and strength in me as a child, all because of a group of motivated, dedicated people who knew what to do.

Though the impact they made on her life is beyond comprehension, these people have no idea who Mrs. Rosario is. To them, that simply does not matter. They are doing this kind of work for hundreds of patients every day. Learning from their example motivated me to deepen my focus on my own purpose. It showed me that when we all live our own purpose-driven lives and really listen to those around us, our actions and interactions are capable of coming back full circle. The positivity that results is fruitful beyond measure.

We all have a duty to pay attention to the signals and trust the decision-making system. When faced with a challenge outside your immediate control, you have to be able to make decisive decisions, know exactly whom to call, and picture the outcome. After all, every organization should ultimately be a community with a set of shared values and goals. When the decision-making system is consistently well managed and based on objective criteria, the idea meritocracy works and can be very outcome driven, as was the case for Mrs. Rosario. Of course, the outcome will not always be as perfect and happy as this one, but no matter what, it most likely will be important.

RED FLAGS

Unfortunately, not every interaction with a professional or expert is going to be inspiring. There are certain things to look out for when seeking outside consultation while making sure you get the best fit for you, your purpose, and your organization.

RÉSUMÉ BUILDERS VS. VALUE ADDERS

Sometimes, the stars align. You are searching for the perfect individual when suddenly, their résumé comes across your desk. On paper, they are everything you need. But not long after their hiring, you realize they are not contributing to the extent you expected them to. They are very positive, which is great, but you get the sense they are creating a false sense of optimism and may not be giving you a realistic picture of what is going on or what needs to happen.

I call these people résumé builders. They are less interested in supporting your purpose and more interested in padding their résumés. In my experience, all they want to do is superficially solve one problem and move on to the next thing instead of getting to the root of what is happening at your organization. They are basically giving you useless information—nothing of value you would ever base a decision on.

By contrast, value adders are people like Dr. Kohrt. His main concern was not adding another achievement to his extensive list of them or even compensation. He was selfless, wanting to help a population who did not have any other options, and he knew he could be a conduit to help get the necessary technology to those patients. His expertise was what was required in order to affect that change. Maybe we would have been able to figure it out ourselves eventually, but he lit a fire under us. In addition to helping with the final clinical trial design, he was also an important deciding factor that convinced me to collaborate with another value adder, who ended up being extremely instrumental in our success.

I had been considering hiring Dr. Sharron Gargosky as my chief of clinical operations based on another recommendation, but I was hesitant because she was not entirely open to relocating to San Diego, where we were based. I was certain there was no way we could run a mega clinical trial without her leading the clinical team at our headquarters full time, but Dr. Kohrt made me consider otherwise. "If Sharron is interested in working with you and she is willing to sign up, just sign her up," he said. "Where she works will not be an issue."

To me, that was cogent advice. I ended up hiring Dr. Gargosky. She quickly became one of the most effective operations persons I have ever had the privilege of working with. She understood the type of organization we were trying to build and, within days, had mapped out a plan for us to meet our goals. She did whatever was necessary for the vested interest and success of the entire team. She effectively managed when she was out of the office, establishing consistency with her team on everything from the cadence of meetings to ensuring week-over-week progress. She made a point to regularly commute to San Diego about four times a month to stay on top of things. She was even successful in spearheading critical on-the-ground efforts for our melanoma trial globally. Because people like Dr. Gargosky are so good at what they do, they have an innate sense of self-accountability and ownership around their work. As a leader, you must trust that. When you find the right person, you have to relinquish some control, believe in their commitment to your vision, and let them do what they do best.

It was auspicious to have Dr. Kohrt's endorsement of Dr. Gargosky. Value-adding experts have the ability to become the heart of an organization and carry the momentum necessary to help the team across the finish line. The anchor of a relay team usually carries the fastest split, and the combined effort of the team is usually faster than if one person tried to do the distance alone.

ASSESSING THE SITUATION

I often rely on personality assessments when hiring someone to work for my organization. They are valuable tools that help you get a good sense of what a person is about and how they will fit into your organization's culture. It is not necessarily a complete picture, but it helps to at least gauge what the person's abilities are, their preferences, and their style. The data is objective and probably even more reliable than an interview. We have all been in situations where someone nails the interview and, within their first week of work, turns out to be a completely different person.

When hiring, it is also key to remember that people tend to pick people with whom they can identify. I am extremely guilty of this—early on, when I used to hire people, I usually hired the person I thought I would most likely get along best with. But over time, I have learned there is much more to it. You have to hire the person who is capable of helping you achieve your goals. It helps to have a group of interviewers who can help you with that criteria. For example, whenever I hired a key position, I always had my executive team help me. They know my faults and where the company or the team need the support, and they are great at finding the person who will help me in those areas.

Look for people who are willing to see themselves objectively. People who think they walk on water cannot see themselves for who they really are, and it can be challenging to work with such people. However, when everyone can see their strengths and weaknesses (and are genuinely interested in improving their weaknesses), you create an environment where people are encouraged to learn from their mistakes. As hard as it may be, you also have to assume most people are not going to change unless there is a very clear willingness in their character to do so. It is better to bet on the people who have already shown they can change, rather than hoping for it.

LACK OF EXPERTISE VS. INEXPERIENCE

There is a difference between a lack of expertise and inexperience. Often, we think we lack expertise in something but it is more likely that we are simply inexperienced at it. We all have to start somewhere, from learning to walk or even learning the multiplication table, and this naturally progresses to the point where we have amassed quite the repertoire of skills. It is easy to self-evaluate your experience level over a period of time. In athletics, life or business, you do not have to be the best something when you start doing it, but as soon as you start doing it, you have an opportunity to grow. As Corporate Athletes, there are many things we have learned and become better at during our careers. If you love your job, you will invest time in it, eventually feeling confident as an expert or key opinion leader.

On the same note, it is also important to acknowledge things in your life that you have expertise in but not passion for. Here, it is critical to ask yourself the question: *Why don't I have a passion?* As a Corporate Athlete, this may be the opportunity to add meaning to it. There are many skills that you have and if you added a bit of meaning and purpose to it, you may find renewed use for it. As a former corporate finance person, at this point in my career, it is abundantly clear that I have a disdain for creating complex spreadsheets. But surprisingly enough, when it comes to budget cycles every year, I realize how fully invested I am in it and how interested I am in the different scenarios that support vital decisions for the team. I surprise myself with the satisfaction of what initially feels like a mundane exercise of finalizing annual corporate budgets. The task may appear boring, but I still have that underlying passion and interest in it. I can see how it brings value to the work I am doing, and I have a greater appreciation of new technologies and software that simplify the act of populating cells on a spreadsheet. You may be underestimating how powerful your existing expertise is and you may have strengths that are underutilized by your current job that could be well utilized by someone else if this passion is activated. Try to find the reasons why you do what you do, even in the everyday activities that appear meaningless. I guarantee that if you look a little closer, you will be able to find a vested interest underlining the things you do that collectively lead to your bigger, more heart-stopping goals.

BUILD BRIDGES

When you are feeling short on the expertise curve, I encourage you to listen to people who disagree with you to help you gain the confidence for decision-making. Listening to people who do not automatically share your point of view gives you an opportunity to understand a different line of reasoning you might not have considered otherwise. Once you have achieved a modest sense of success to help gain confidence, it is easy to be surrounded by people who constantly say yes, even when they do not mean it. By frequently challenging your perspective, you are repeatedly asking yourself to grow. Having open-minded conversations with experts or peers is the quickest way to not only further educate yourself, but also increase your probability of making the right decision. It is important to remember that there are many inexperienced people who have great ideas. They too likely have a different perspective and can enhance or offer a "plus" idea. Do not be quick to write anybody off—listen carefully, use your intuition, and make considerations as you go.

Finally, as a leader and world-class Corporate Athlete, it is crucial that you establish a culture where open communication is the norm to help make the conversion from lack of expertise to experienced expert, regardless of position or stature. Do not fall into the trap of underestimating expertise for lack of experience—this is an opportunity for growth to share knowledge, experience, and perspective.

Many moons ago, giving an employee the opportunity to speak with the CEO might have been something special. I, however, believe this is the last thing we need to be worried about. Titles should not make a difference between who you are or are not allowed to speak to. My daughter owns a T-shirt that says *Build bridges, not walls*. Organizational health has to transcend hierarchy.

□

WORLD-CLASS CODA

- In order for leadership to be effective and successful, ego needs to be removed from the equation. It is imperative to keep your eye on the ultimate goal but your bigger priority is to serve the greater community and to elevate those around you.

- Asking for help and also providing help are both sides of the same coin, and they are both integral to great leadership. Reciprocity is often looked down upon in business when people are looking to get ahead. Lead by example and show how both actions can benefit all parties involved.

- Cultivating a workplace culture where expertise is shared encourages the contribution of employees with varied work experiences. Everyone has something valuable to add, regardless of age, tenure or perceived experience. Stay open and you may be rewarded with a new perspective.

CATAPULT FORWARD

- If you were to place a blank piece of paper on each of your team member's desks with a request for feedback on how you could be more effective, what would they say? Do you have the courage and humility to take their feedback to heart, to look inside yourself and make the necessary adjustments, to relinquish egotistical behaviour, and to serve the greater good for your purpose?

- Why do you do the things you do? Can you think of instances when the outcome was favourable because your intentions were aligned? Have there been times where you decided to do something and it did not feel right? How did this play out?

- Who is your sounding board for decisions and ideas? Can you think of three people in your life who have been instrumental in assisting you in your career? What qualities do they have in common and how do you emulate this today in order to pass the kindness on to your colleagues or mentees?

Joy is a great motivator. It is the one quality that will pull you out of the inevitable failures you will encounter, setting you back on course and providing the much-needed inspiration you will need to keep going. Joy is a harbinger for success. When you find it, harness it and use it as a vehicle to achieve your goals. Putting your purpose to work absolutely involves a certain amount of play—nurture that spark and see where it can take you!

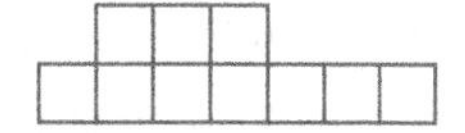

PRINCIPLE 9

JOY

"Stay hungry. Stay foolish."
—STEVE JOBS

Saving this principle for the end of this book is not an indication that it is of lesser importance because without joy, why do anything at all? Joy is defined as a feeling of great pleasure, evoked by well-being, success, and an overall state of happiness. It is absolutely integral to your success as a Corporate Athlete.

I am lucky enough to be surrounded by an incredible group of friends and family. Some are entrepreneurs, some are professionals, some are athletes, and others are creative in their own disciplines. They all have an incredible sense of humour, happiness, and joy. They are successful not by measure of their physical wealth but because they live life fully. In different discussions during the research of this book, there was a resounding consensus that the happiness and joy they experience from their work is derived by focusing on their passion as opposed to financial motivation.

I have always been fascinated by people who have the ability to turn their passion into a vocation. Their love for what they do informs and evolves what they do—there is room to play, think creatively, and execute fresh and novel ideas. Enjoying what you do changes your approach to your career and life. In the midst of pursuing education, higher learning or your work life, joy often gets pushed to the side and forgotten over time in favour of goals society and academia deem to be more important.

When I was young, I thought I would have my career and then outside of my career, I would find joy. In my mind, work and joy did not intersect. I grew up surrounded by immigrant families, my own included, who mostly worked labour jobs just to keep their families afloat. My mom certainly did not find joy as a prep cook and my dad came home exhausted after his shift at the convenience store. Their joy was limited to spending time with family on the weekends.

When I started university, I had this ambition of graduating with a dual degree in business and performing arts and someday becoming an artistic director of a playhouse. I was looking for a way to merge my goals with my hobbies, to bring joy to my career because I did not see this modelled in my elders when I was growing up. Over twenty years later and on a very different career path, I feel that business has allowed me to express my artistic side and achieve that sense of joy I longed for. It sounds contradictory, but as an entrepreneur, I have a blank canvas I can use to push the boundaries of expressiveness. There is an endless ingenuity that comes with a creative mindset, often leading to innovative breakthroughs. As a Corporate Athlete, how will you jump-start the creative process to bring joy to your everyday tasks and to tackle the inevitable challenges in your career?

Working in life sciences has certainly given me a very different perspective on creativity. Through science and R&D, my colleagues teach me every day about reaching beyond perceived limits and how to dream outside the borders of reality. In early-stage research, failure is constant, yet they persevere. It would be unwise for us to overlook this resilience. How do we borrow from similarities in different industries and apply it to careers that may not appear outwardly artistic or creative? How can we push the boundaries and make time to feed our imagination so we can approach our careers in a different way? Opening your mind to new ideas does not just happen by chance or by luck. It takes work and commitment. Establishing a personal structure and a dedication to it gives us the freedom to expand, opening us up to a world of possibilities. Trust me—it will be worth the investment.

FIND YOUR JOY QUOTIENT

Work-life balance is a lovely notion. It may even work beautifully for some people. But if you are a person who strives for more and who is interested in making a real impact, the societal norm of a work-life balance is often inconsistent with living in purpose and the life you have chosen.

I suggest we focus on work-self-life passion. Work-self-life passion is about experiencing enthusiasm in all aspects of your life while managing the expectations. It allows you to both enjoy your work and have time for the things you care about the most, including the people and personal causes you hold most dear. It is not easy for a Corporate Athlete to base their success on how productive they are between 9 a.m. and 5 p.m., then switch gears and become the attentive family person they need to be at home. If you are trying to live a balanced life while chasing your goals, you will also be tortured by feelings of guilt and inadequacy because you will be falling short in every role. You must stop putting boundaries around yourself, embrace who you are, and find the balance that allows you to be that person in every area of your life so you can reach all of your goals. You need to reframe how you think about balance in order to allow yourself to reach it.

I encourage you to break it down and think clearly about the distinct categories in your life:

- **Work**—Everything related to your job, including social events and travel. Work includes every single thing in your life related to bringing home a paycheck.

- **Self**—Things you do just for yourself. This includes anything related to self-care—exercising, meditating, reading—anything you do for you, and you alone.

- **Life**—The people and causes that matter to you outside of work. These are things you care most about that have nothing to do with furthering your business interests.

Now, imagine you can quantify all of this into a currency for joy. This becomes your Joy Quotient. Just as you would save money to invest in the things you want, you can have a currency to spend on the joy you want in your life. For example, my friend Winnie has developed a similar concept she calls a "love bank," based on the idea that in relationships, you can invest units to draw upon later. Couples often have to make sacrifices for each other and support each other in different situations. Having a healthy balance in the love bank helps in those times. The Joy Quotient is exactly the same—do you consider time spent in these different categories of work-self-life an investment or an expense?

To determine your Joy Quotient, think about the value you assign the areas of work, self, and life. Determine and assign a percentage out of 100 to each category based on what your ideal split would be. Finally, think about the things you need to either start or stop doing in order to align with your ideal Joy Quotient. What are the things you will choose to do in order to achieve these determinations? The following grid can be a helpful tool to get started:

	WORK	SELF	LIFE
CURRENT JOY QUOTIENT			
IDEAL JOY QUOTIENT			
START OR DO MORE OFTEN			
STOP OR DO LESS OFTEN			

Joy Quotient

Ideally, reaching your Joy Quotient is reaching personal satisfaction and the happiness it brings. The idea is to be able to make adjustments in your life until it feels right for you. The table above may help guide you until you get into the habit of doing this exercise regularly. I encourage you to check in with yourself periodically, especially if you start to feel stressed or if you find that sense of joy is missing.

The questions at each check-in then become: Where is your currency being expended? Are you only investing passion at work and you and your family get the scraps leftover at the end of the week or vice versa? Is your work suffering because you have a new person in your life or someone or something else is demanding more of your time?

When you take the time to sit, determine, and calibrate your ideal Joy Quotient, you will find your sweet spot for living in a way that is completely aligned with your purpose.

STRESS

When you have a system with components working in harmony like in the CAHPT, stress is the wrench that can disrupt the entire complex. Stress is the burdensome state of body overload—we have all felt it. It does not feel good to be rushed, hassled, anxious, abandoned or apprehensive. I discuss the positives of physical stress within limitation throughout athletic examples in this book. However, emotional stress overloads circuits. Excessive prolonged mental stress becomes emotional distress. In time, excessive levels of stress create a breakdown in the body's self-defence mechanisms, which is why it is often cited as an underlying reason for many diseases.

BREAKAWAY

KEEP TWEAKING

Investing in your sweet spot is not something you can do once and quit. When you are doing the things you love, you are also delivering the most value. Therefore, continuing to push down on that gas pedal is crucial. Michael Jordan did not realize his talents on the basketball court and then stopped practicing. Roger Federer did not become great at tennis and consequently stopped investing the time necessary to become even better at it. When you find your sweet spot, you have to continue to build that resilience, so mediocrity is no longer a word in your vocabulary. When you are pushing yourself to the best of your ability, experiencing success and recovering from setbacks, you are more capable of bouncing back when you encounter challenges in the future.

Some people are comfortable never fully tapping into their true purpose, settling on being a jack-of-all-trades. I always encourage members of my staff to focus on a specialty because when they do, they become irreplaceable. Sometimes, especially at small start-ups, wearing many hats is part of growing and gaining experience but, eventually, you have to make sure everyone's job is maximizing their particular skill set.

When you do recognize your strengths, it becomes your responsibility to ensure you continue to grow them. After my first Ironman, I was completely burnt out on training. I basically hung up my bike and I did not even want to look at it. I was still swimming and doing some running, but because I also was still eating like I did when I was training, I ended up putting on twenty pounds. I was in a rut because I was out of balance. I had so much structure going into that race and afterward, I threw my hands in the air and took an entire year off.

That first race was supposed to be a one-off, something to check off my bucket list. But then, I got the itch to do it again. I used that second Ironman as motivation to put the structure back in my life. This time, I was very cautious about avoiding the same "drop off a cliff" feeling I had experienced after the first race. I adjusted my expectations so afterward I could maintain most of the structure I had put in place on an ongoing basis.

After my second Ironman, I only took a week off to recover and since then, I have been training consistently. I am not training at the exact same level I would be in the weeks leading up to a race, but I am maintaining my level of performance. There have been times when life has thrown me off because of travel or my schedule but, overall, I feel as though I have become truly comfortable with a steady training regime that allows me to have balance without ever feeling burned out.

Now, I am in my forties and feel like I am in my best shape ever. I am running and cycling faster than I ever have. My 10K and 15K times are the fastest I have ever recorded. I owe this to the fact that I did not let myself stop simply because I had crossed yet another finish line.

If we rewind a bit, it might help to look at the CAHPT again from a different lens, based on the following equation:

IDEAL PERFORMANCE STATE (IPS)

=

HEALTH & WELLNESS

+

OPTIMAL RESPONSE TO ENVIRONMENTAL STRESS

IPS Equation

IPS is achieved when you are able to create a balance. We have already discussed how important health and wellness is in all aspects of your life, but it is especially imperative when you are a Corporate Athlete. In addition to taking care of our own health, we also need to be able to deal with environmental changes. Stress can impact your IPS by breaking down your perfect plans. The Corporate Athlete is constantly working under stress because we are learning resilience, but we are also human. Stress leads to a breakdown of the IPS and therefore our life balance, throwing off our Joy Quotient.

RESPONDING TO YOUR ENVIRONMENT

Some days, I sit at my desk for fourteen hours. I have call after call, meetings, fires to put out, and in between, I have my actual tasks. Those days, when things do not go as planned, I use my toolkit:

Meditate—This term gets tossed around lightly, but I believe it is your most important tool. Whether you are a novice or seasoned meditator, making space for five to ten minutes of meditation time can improve your day.

Write—Grab a journal and write down a few lines. Venting on paper or putting your feelings into words allows you to have an outlet so you can be more mindful of your interactions and reactions with your colleagues and employees.

Stretch—Sitting in front of our computers all day is a recipe for bad posture and holding your breath during moments of stress is not ideal for anyone. Get out of your seat, stretch, and take a deep breath. Even short breaks and a burst of movement can be a much-needed reset.

Workout—If you are able to make the time, go sweat it out. Run, bike, walk, and leave your phone at home. Disconnect and make this time about you.

Speak—Phone a friend and make a connection. Borrow and lend an ear.

Once you have had time to regroup, use this opportunity to revisit your personal CAHPT. Decide if you need to recalibrate in the presence of a changing environment. Life happens—there may be changes at work or at home, circumstances outside of your control. Use the tools you have to get yourself back on track.

LIFESTYLE VS. IDEAL JOY QUOTIENT

An important part of finding work-self-life balance is redefining the concept of "lifestyle." Rather than aspire toward a lifestyle you think is ideal, look inward and make it personal. You can create the lifestyle you want for yourself by aligning it with your ideal Joy Quotient.

For many people, lifestyle has become about emulating a certain way of life, specifically the ones associated with celebrities. We see what is presented on social media and assume people are experiencing happiness at all times. But everyone has their own Joy Quotient, and we have no way of knowing if those people are effectively investing in theirs. Chasing a lifestyle that does not align with your purpose can cause a great deal of stress and anxiety, and even if you achieve it, the results might not be what you expect. People who strive for wealth and status for the wrong reasons are nearly always dissatisfied. Each individual must look within to become confident and comfortable with their own Joy Quotient and s*top measuring it against others.*

There are many different indicators you can use to measure your own unique Joy Quotient. How satisfied are you? Are you enjoying your workout, or does it feel like a chore? Is family time lacking this week because you had deadlines at work and now you are feeling guilty for spending less time with the kids? Find a way to balance things again. Change your workout to something that is more satisfying. Apologize to the kids for not being there and do something fun to make up for it when the time allows. Most of all, have something to look forward to.

An effective way to guarantee satisfaction is to incorporate something you look forward to into your regular routine. For me, that is my sundae before Monday. I make it a point to have an ice cream sundae every Sunday evening. That might not seem like such a big deal to you, but there is a lot more going on than indulging in a weekly treat. That sundae:

- Gives me something to look forward to;
- Reinforces the structure of my routine;
- Puts me in a good mood, setting me up for a happier mindset when I wake the following morning so I can attack my Monday;
- Is the inspiration behind my family nickname, DJ Hard Shell, a nod to my favourite topping.

No matter how crazy the week is, I always know that moment is waiting for me and when it comes, I am going to deserve it. This tradition first started on Friday nights with my friend Jacob, our nanny's boyfriend, who became a close family friend. We would make these epic sundaes at the end of the weekend with all kinds of toppings, unwinding and laughing about the week. It was such a simple addition to my routine, but it became a new way of measuring my Joy Quotient. Most people dread Sunday evenings because they know Monday is on the horizon. My Sunday nights make my Mondays that much sweeter. My weekly sundae is now just another way for me to check in and make sure I am making time for myself. For you, it might be a regular date night or a daily walk. Whatever it is, it is something that helps you keep yourself in a routine and brings you joy, even if the week ahead is looking anything but joyful.

When you are determining your ideal Joy Quotient, you are a lot like Goldilocks trying to figure out what the ideal temperature for porridge is. Budgeting a great deal of time and energy on one thing might be too much; ignoring it completely is also not going to work. You have to find the spot where everything is just right. After all, each aspect of your Joy Quotient is made up of micro purposes that when combined and working together in harmony, enrich and support your macro purpose. The Joy Quotient is not static—it needs to have room for movement, and you have to be willing to adapt to various scenarios.

It is important to remember your Joy Quotient is inextricably linked to purpose, and purpose is linked to your drive to get things done. Therefore, your Joy Quotient is not about overdoing it in any one area. If you compare it to finances, it is about stashing away however much you need in each area in order to have a healthy reserve should you need to tap into it. Creating that reserve will get easier over time and the best thing you can do is to start saving early. You will soon realize that working on your Joy Quotient will get easier and more fluid over time.

WORK-SELF-LIFE AND PERFORMANCE IN THE WORKPLACE

Your Joy Quotient drives performance; in turn, good performance, which depends on consistent progress, enhances work-self-life.

One of the primary things that will ensure your Joy Quotient in the workplace stays high is health and wellness. Personally, I like to get people up and moving every hour on the hour in the workplace. My colleagues know that when the clock strikes noon, we are all going to jump up and do ten push-ups or squats together. Or, if you are not inclined to do something as physically challenging, you can simply take ten deep breaths. It is an easy way to get people to take little breaks to break up their day and keep up the momentum—think of it as a collective breath.

As a leader, you have the responsibility to help ensure the workplace has a high Joy Quotient. To be clear, it is not your responsibility that your employees have a high Joy Quotient in their life, but you can certainly help the work-self-life balance from a work perspective.

For example, during one of my company's bigger projects, we had researchers, doctors, accountants, lawyers, and clinical operations all working together. We decided it would benefit everyone and the company to have everyone working in the same building to foster connectedness. We were very lucky to have the chance to design and construct a new building with many collaborative workspaces, and with everyone physically closer to one another, real teamwork began. Everyone was able to see how their contribution directly affected other aspects of the project, and productivity was redefined. Instead of having many clusters of people doing different things and then piecing it all together, we created one cohesive unit to reinforce it.

Remember, a fun workplace will not necessarily translate to a great workplace. Fun is not a driver of a great workplaces, but it is an important barometer. Having fun at work does not mean getting a cake and a keg and forcing people to party. Fun is when employees are enjoying their work and feeling supported and encouraged. Take the time to recognize organizational milestones as well as those happening in the lives of your colleagues because it will go a long way, both in terms of showing you care and in bolstering your own Joy Quotient.

TAKING BACK LUNCH

The lunch hour has become a hurried gobble of a quick sandwich while working. I learned this the hard way, as the norm of eating at your desk is very much a characteristic of American culture. When I worked in Canada at the law firm, we took lunch. Things were busy, sure, but we nearly always had time to take a step back and socialize with one another for a little while. I have found that in the US, people tend to ignore lunch unless they purposely try to make a point of having it. When I first started working in America, I ate the same sad turkey sandwich every day in front of my computer in my dingy office. I did that because it was what everyone else was doing and found it to be quite miserable.

Research conducted by Right Management shows only one in five people steps away for a midday meal[36]. The average worker takes less than

36 "Survey Reveals One-Third of Employees Lunch at Their Desk,"...

twenty minutes away from their desk for lunch. Many never leave their desk at all or get pulled into a lunch meeting or eat while catching up on email or hurrying to complete a task. Tony Schwartz, author of *The Way We're Working Isn't Working* and CEO of the Energy Project, wants to change that.

Schwartz, in an initiative called "Take Back Your Lunch," suggests workers reclaim their midday break for the sake of health and sanity, not to mention their productivity. "The demand in people's lives overwhelms their capacity. We need to stop operating as if we were computers—we operate better when we pulse between spending and recovering energy," Schwartz says.[37] A lunch break is not only important from a nutrition standpoint to maintain stable blood sugar levels and renew energy, but it also gives your brain a chance to recuperate.

When I started my own company, I made sure people knew they could disengage for lunch. I would have lunches provided a few times a week to encourage everyone to come together, take a little break, and interact. The connections that can happen during lunch are fundamental for joy overall, because otherwise, you can go to the office, only interact with certain people for specific reasons, and forget about the personal side of the cause you are all working to help advance.

...*Employment Law Daily*, https://www.right.com, https://www.employmentlawdaily.com/index.php/news/%20survey-reveals-one-third-of-employees-lunch-at-their-desk/ (accessed on April 19, 2020).

37 Courtney Hutchinson, "Lunch Hour Love: Workers of the World, Grab a Bite!" *ABC News*, June 25, 2010, abcnews.go.com/Health/MindMoodNews/back-lunch-hour-back-energy/story?id=11012804 (accessed on April 19, 2020).

THE TRICKLE-DOWN EFFECT

Once you open your life to more joy, you will soon see how positivity begets positivity. The clearest example of this is how the environment you create at home and in your personal life directly influences your work life. People and their energy, both at home and at work, can lift you up or drag you down. I am fortunate enough to have a person in my life who excels at helping me keep things in perspective and focus on the positive.

My wife, Nina, is dedicated to making sure our work-self-life balance is level. Because of my career, I have spent at least 50 percent of my time away from my wife ever since we met. I board a plane more than 150 times a year (I keep count). When I am home, Nina expects me to be present. We are deliberate about spending time together. We seek out fun things to do together as a family. Excessive email checking is not tolerated, and she can proudly say that she once walked out on me at dinner because I was on my phone.

All of this is not only for her sanity, but also for mine as well. Early on, Nina recognized that structure gives me a stronger sense of purpose. Her expectations of me when I am home actually help me be more productive at work. How can I be effective if I am trying to be everything to everyone all the time? Setting boundaries helps me show respect to my loved ones by encouraging me to give them all my attention and focus when I am home.

That is not to say she does not have to rein me in from time to time. At the time of this writing, I am approaching my fortieth birthday and I suddenly feel as if I have to accomplish everything. I keep signing up for stuff I have no reason to be signing up for. It has come to the point where Nina said, "I need you to write down your three-month plan because the list of things you are trying to do is getting ridiculous." That plan also includes how family factors into all of the things I want to do because while I know in my heart that they are always my top priority, it is possible to lose sight of that fact when you are weighing yourself down with too many distractions.

At the same time, my family does recognize that as a self-employed individual, I do have to hustle, and they let me go when I need them to. But because of Nina's influence, I never lose sight of the big picture and what really matters.

We have a date every Friday around 6 p.m. I could be working on the deal of the century but if that time rolls around and I am making no obvious attempts to stop working, she will simply say, "If you do not get off that phone, we are not going out." To this day, it has never hurt me or my career to tell a person, "I cannot speak to you right now. We will pick up this conversation tomorrow." These days, people are beginning to have more respect for the time and boundaries of others. They are not interested in hijacking somebody's time just for the sake of doing so. Prior to the COVID-19 pandemic, I found it easier to call a colleague on a Wednesday evening after dinner to chat about work. Now, I am also aware that with kids attending virtual school or after-school activities being turned on their heads, everyone has made adjustments to their time commitments. I have also found myself on the receiving end of a, "I have to go; we will talk later," comment, and I actually find it makes me respect the other person even more. It shows me they have their priorities in order and are adhering to a structure they have created for themselves. I know firsthand how having a supportive structure underneath you can reinforce your purpose. No one is shortchanging anyone else—they are making choices that will help them be as impactful as possible in all areas of their lives.

BREAKAWAY

THE COFFEE RIDE

To me, bike riding is usually about training. I have always approached my outdoor rides (or any cycling for that matter) as a training ride. I become very focused on doing a certain amount of mileage in a certain amount of time, which is easy to do when you are riding solo. My group rides usually involve my Breakaway team and, because I am competitive, I see those as challenges to keep up with the strongest riders. For a long time, cycling was never about a social ride for me.

That all changed last year when I started riding in Vancouver with a bunch of buddies. The group included all levels of riders, but it was not about competing. It was about laughter. It was about fun.

On every ride, we were always looking for the best place to have coffee and a chocolate chip cookie. Every route had to include a spot that served very good coffee and awesome cookies. I never would have thought of doing that on my own, especially not when I am training with the team in San Diego because that team does not stop for anything, least of all cookies. That is not what that team is about, and that is fine. But from this other group, I learned that the act of injecting joy into your training can result in a shift in mindset. You still end up doing the same distance, the same time, putting in the same effort, but you forget about any of the pain associated with it because you *just had so much fun* along the way.

As a Corporate Athlete, you also can benefit from infusing joy into your work by surrounding yourself with people who are motivating you and making it easier to reach your goals. You need to be around the people who bring the best out of you. Then, when you are tackling that daunting distance or navigating an unforeseen challenge, you are able to still have fun and that will reinforce your purpose and keep you moving, especially when the road gets rough.

LAUGH A LITTLE (OR A LOT)

If entrepreneurship requires you to work harder than you have ever imagined, remembering to laugh along the way will make things easier. There will be times when you will have to push yourself to the limit. Sometimes, even just the day-to-day grind will be a challenge. In order to maintain the level of energy and endurance you will need to continue to live a purpose-driven life through all of this, you will need one crucial element: laughter.

Sometimes handling conflict well requires the use of humour, as opposed to falling back on defensiveness or reactivity. The use of humour relieves tension and, at the same time, promotes a collaborative spirit by making the situation fun. Most of the time, it is as simple as reframing an idea or comment or emphasizing the excitement of fast-paced competition rather than the stress of competing in brutally tough and uncertain markets. Laughter is an important part of the entire day for me, from the first interaction with Nina and my daughters to management meetings.

While I do not suggest using humour to convey negative news, it certainly depends on the negative news. The point I want to make is that you can use tact with a little levity to say things that might otherwise give offence because the message is simultaneously serious and not serious. The recipient is allowed to save face by receiving the serious message while appearing not to do so. The result is the communication of difficult information in a more considerate and less personally threatening way.

Humour can also move decision-making into a collaborative rather than competitive frame through its powerful effect on mood. There is a fairly large body of research that claims people in a positive mood tend to be not only optimistic but also more forgiving of others and creative in seeking solutions.[38] A positive mood triggers a more accurate perception of others' arguments because people in a good mood tend to relax their defensive barriers and are able to listen more effectively.

Seeking out opportunities to add humour and fun will give you a positivity that fuels your inner drive. Without it, you risk losing your motivation. Your ability to achieve your purpose becomes impaired and you stop finding meaning in your work. You start hoarding your energy and dedication, pouring it all into something you no longer find fulfilling rather than giving it freely to the things that bring you joy. You stop inspiring others because your focus is only on yourself.

Keep in mind that this does not require a herculean effort. Even small changes can make a huge impact once you realize how much of a difference a little levity and positivity can make.

THE POWER OF 'NO'

You are entitled to say no to what gets in the way of your purpose and what blocks you from a life of abundance. In order for you to have true accountability in your purpose and your unique contribution to this world, you will likely have to use this word occasionally. Saying no when people or things are going to negatively affect your Joy Quotient is absolutely okay. You cannot do everything for everyone. Most times, we have to say no when something is going to negatively affect the work-self-life balance. It is not worth it to say yes out of pressure or obligation. Unless you have ample room for it, do not force yourself into that deficit.

---

38 Barbara L. Fredrickson, "The Broaden-and-Build Theory of Positive Emotions," *Philosophical Transactions of the Royal Society of London. Series B: Biological Sciences*, 2004, vol. 359, no. 1449, pp. 1367–1377, doi:10.1098/rstb.2004.1512.

BREAKAWAY

THE JOY IN PAIN

I like the sensation of pain. To me, it is a way of measuring my performance. My family knows that when they hear me yelling in the morning, it just means I am on the treadmill, screaming to motivate myself. The scientific term for attempting to establish a sense of self-control over pain is known as adaptive coping strategies. I believe humour is a very effective example of this mechanism.

Having completed various marathons and an Ironman, it is no secret that I have experienced an excruciating amount of pain. Normally, I am able to laugh at the pain. Most of the time in endurance training or competing, it is not really about perceived pain, but rather, it is about getting comfortable with the discomfort for extended periods of time, something many of us tend not to do very often to our bodies. Part of endurance training is centred around getting comfortable for a period of time with the training hovering near the threshold of your work capacity to maintain the benefits while also staying under the threshold to allow your heart to keep pumping oxygen-rich blood to all the moving parts that need to perform. We touched on this concept with the work capacity sink analogy in Principle Seven, which focuses on the zone of discomfort and happens to be a large part of building resilience to improve your capacity.

Similarly, Corporate Athletes have to endure pain in order to grow. You will have to fire people you do not necessarily want to see go. You will confront uncertainty. You will deal with adversity. As you do, you will recognize all of it as the good kind of pain, the pain you learn from and emerge on the other side of as a stronger leader.

The pain you experience is just another opportunity to build your resilience. Painful moments—whether they are physical, mental or emotional—give you the experience to deal with whatever is next. It will make you better equipped to deal with stressful situations and will give you the confidence and courage to extend beyond your comfort zone, encouraging others around you to do the same.

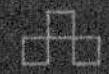

COMMIT: A METHODOLOGY TO THE MADNESS

We have all had those nights: too much food and wine with great friends and even greater conversations. The next day, I woke up at the break of dawn to get ready for my Sunday run, still feeling hazy, my legs still stiff and sore from the ride the day before. I really could use a few more hours of sleep. But a few minutes later, I am already in the shower, contemplating the day. I put on my running gear, hydrate, grab my sports gel, and head out the door. On this particular morning, this routine is a drag because my bed was so comfortable. After a short five-minute jog, I make it to the trailhead of Los Penasquitos Canyon Trail and at that exact moment, I am reminded why it is all worth it. The endless rolling hills of mixed landscape of the Southern California desert preserve, the sun beginning to rise, the crisp air, and nobody in sight. I start my run, a seven-mile loop, my legs starting to pick up the pace, my mind and body beginning to connect. As I make it past the first mile, the thought of my comfortable bed is replaced with the pure joy of being outside.

We must deliberately make the commitment to whatever discipline or decision we have chosen. There is no single structure that brings joy to everyone, but everyone has a Joy Quotient. In order for your individual process as a world-class Corporate Athlete to work, day in and day out, over weeks, months, and years, there must be a structure supporting the results you seek. Creativity balances structure.

The structural part of the Corporate Athlete High-Performance Trapezium and the principles throughout this book are what transform you from a one-hit wonder, ordinary worker or hobbyist into a high-performing disruptor. Creativity dislikes the mundane part of the process: the work and the discipline; the thoughtful, measured calculations; the endless tedium of organizing; the structured and operational part of anything you are trying to achieve. You need the ideal Joy Quotient for both parts to work—the creative imagination and the structure to hang it on. Yin and yang. With both, there is no doubt you will get there.

□

WORLD-CLASS CODA

- Find joy and an overall spirit of positivity in your endeavours and use these qualities as motivation to refine your social and professional resilience.

- Money cannot be the be-all-end-all. Redefine what success looks like to you and aim for the sweet spot where you are comfortable yet still inspired to do more and do better. This "hustle zone" will allow you to remain creative in order to achieve your goals.

- Try to stick to your training program even if you are travelling. This forces you to check in with your body, enabling you to develop greater kinesthetic awareness and encourages continued learning. When you constantly challenge yourself, you will be rewarded with greater proficiency, allowing you to reach your personal bests.

CATAPULT FORWARD

- What is your ideal Joy Quotient? Do you have adequate time to yourself? What are your rules for work, self, and life? How is this contributing to your management of stress?

- How is your work environment affecting your well-being, engagement, and productivity? Is it building resilience?

- How can you establish a system in your life to find a way to reward yourself with the appropriate time to adequately recover and renew? How can you ensure commitment to your routine even when the timing feels less than ideal?

The world requires people who are willing to challenge the current state of affairs. There is always room for improvement and your willingness and desire to disrupt the status quo in the pursuit of global change is desperately needed. Every success story begins with one idea from one person that instigates a movement to create big changes. Be courageous and use this fearlessness to make your mark on the world. We are waiting for you.

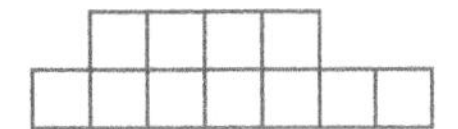

PRINCIPLE 10

MAVERICKS

"No one has ever become poor by giving."
—ANNE FRANK

Not all races lead to a podium finish. Not all projects are winners either. When you have a plan in mind, every training session and project begins with promise and hope. We push on, believing we are building toward something amazing.

When I am working on a new project or a new business, I will file hundreds of initial ideas. Over the course of a month, about thirty to forty different objectives will be pared down to no more than five that will set the corporate strategy and be used to inform the different stakeholders, driving accountability for myself and the entire team. Once those corporate milestones are set, I will not look at that document again for a month. Rather, during this time, I will work with my team to institute the necessary underlying detailed objectives to the goals and our structure to ensure that our milestones can be met. After every quarter, we measure against our progress. When I am preparing for a race, it is no different. For a marathon, the training plan encompasses approximately the four months leading up to a race and if I am getting ready for a longer endurance event like an Ironman, it can begin up to five months before a race. During these days, weeks, and months, there are inevitably ups and downs. In both cases, I often feel like I am running short on time and patience. But after the initial corporate board package is released, it is out of my hands. After all the training is completed and I begin the taper, my primary focus is to relax, avoiding any injuries. If I have put in the training, then I will not think about my level of preparedness. Instead, I put on my headphones, listen to one of my favourite playlists, and I reflect. Sometimes, everything is in accordance with the plan and sometimes it is not. Either way, the constant and reliable force is time and in order to be unrelenting, you need to not only take advantage of time, but also put in the time.

Do not expect everything that you touch will turn to gold. The Corporate Athlete journey of resilience will take you through a winding path, through wins and losses. What we experience is change, not time. It is absolutely normal to have projects that fail, training plans that do not work out as planned. Things will come up or get in the way of your meticulous master plan. Failure is part of the adventure. If you are working on implementing big changes, at some point, there will be challenges and setbacks, otherwise everyone would be doing it. The trick is learning how to handle failure when it arrives. Will you wither against it and give up? Or will you get back up time and again? Aristotle observed that time does not exist without change; we also know that success does not exist without failure.

After the inevitable pain-and-torment phase subsides, a Corporate Athlete uses failure as a learning experience. They take all the relevant data from these experiences to learn as much as possible. This is the time to take advantage of. The speed of change and conversion to success is determined by us, similar to how Aristotle defined time as simply the measurement of the difference between "before" and "after."[39] Our experiences are malleable and what we decide to learn from them help to establish the pace for our future success. Today's failures are often the basis for the next big win. Think about the world-class Corporate Athlete mindset as the long game. In it, there are always incredible highs and devastating lows—that is the nature of any physical or business progress over the long term.

Here is some good news. Once you understand your individual world-class Corporate Athlete mindset and the steps that will lead you to your CAHPT pinnacle, you can use that same system over and over again. You set the pace. Once you are able to define your process, it becomes easier. Lather, rinse, repeat. This only works if, at the end of each cycle, you analyze what you have done, learn from it, and modify parts that can be improved upon. This is the essence of thinking like a Corporate Athlete to build resilience. This ability allows the mindset to tackle anything. To become better global citizens, we need to take what we have learned to take on some of the largest challenges on this planet.

39 Andrea Falcon, "Time for Aristotle," *Notre Dame Philosophical Reviews*, April 1, 2006, https://ndpr.nd.edu/news/time-for-aristotle/.

We are all resilient Corporate Athletes who are capable of using that resilience to overcome the challenges that come with big opportunities. We all have a vast, untapped accountability to purpose within us. This is an opportunity to step toward the edge. Your next adventure awaits you.

INSPIRING CHANGE

In today's modern world, we are trained to deal with problems in a certain way. We are constantly surrounded by things that distract our ability to be purpose driven and create greater social value. If you took just one of those distractions—say, nonproductive screen time—and measured the amount of time you spend on it, you could see just how much time you are wasting. If it is even just a few hours a day, over a lifetime, that adds up to years of your valuable time on this earth that you could have been focusing on making the world around you a better place.

We need to detox ourselves and shift our thinking to be clearer, deeper, and more focused around relationships in order to create more social value. Unless we are willing to accept the idea that things can never change, we have to show a certain degree of engagement. We all have different things in our day pulling us in a million different directions. How do we shift that mindset, so we are not always getting so caught up in the day-to-day minutiae of life?

A study published in the journal *Psychological Science* found that we may get longer-lasting happiness by giving to others, rather than receiving for ourselves.[40] The study involved a series of experiments wherein participants

40 Ed O'Brien and Samantha Kassirer, "People Are Slow to Adapt to the Warm Glow of Giving," *Psychological Science*, 2019, vol. 30,2: 193-204, doi:10.1177/0956797618814145.

either gave a gift to themselves or to others to see which brought the longest-lasting joy. In one of the experiments, ninety-six participants received $5 every day for five days and were randomly assigned to spend the money on themselves or on someone else. They could so do by leaving money in a tip jar at a café or making an online donation to charity every day, but they had to spend the money on the exact same thing each time. The findings showed that when they repeatedly gave to others, the participants' happiness did not decline, and the joy they felt from giving to others on the fifth day was just as strong as on the first day. In contrast, participants who spent money on themselves reported a steady decline in happiness over the five-day period.

How do we overcome the defeatist mentality permeating society right now? So many people feel there is no way they can make a difference because the state of the world is so bleak and the challenges standing in the way of change seem insurmountable. It can be tempting to stay stuck in your own world and focus only on taking care of yourself. This is incredibly easy to do in this modern world, with the constant barrage of things being thrown at you all day. But even making small changes on a personal level can cause a ripple effect and in that, there is great power and value. Every minute of the day, you have an opportunity to make a small change and show a level of engagement and interest. It is there in every interaction you have, whether it is with your loved ones, your coworkers or your neighbours.

In early March 2020, most of the world was going about their daily lives. The novel coronavirus was in the news, but for most of us, it was just news, not an eminent threat. Outside of the images we saw, there was little to suggest a global crisis was unfolding. By the end of March, everything had changed—over 90 percent of the world was under some form of lockdown and soon, millions of people were testing positive for the virus. We did not have time to adjust. Change just happened. Even if our lives felt static at home, during that same period, we witnessed rapid acceleration in technology adoption, education disruption, life science cooperation, and even environmental improvements around the world. All of a sudden, video calls, virtual learning, online workouts, and on-demand grocery and takeout delivery became the norm, confronting the reality of why we had not taken advantage of these tools in this way before. Since that period, a new baseline has been set and we have learned that no industry or business operates in a vacuum. Whether we took a moment to reflect on it or not, we just bore witness to an extremely transformative period.

As an entrepreneur, you have to remember that being an agent of change is not about "winning;" it is about having purpose in everything you do. You cannot "win" at improving education. You cannot "win" technology. You cannot "win" life sciences. But as a Corporate Athlete, you can let your purpose dictate your actions and use the motivation it fuels within you to address global challenges.

Change is not constant, but rather, relative. It is the nature of personal life, the workplace, and broader global dynamics. The COVID-19 pandemic revealed and accelerated widespread trends in culture and politics, dynamics that were already present in society. Reflecting back over the last year, there are too many examples of how the recent crisis provided new opportunities—but the more disruptive the crisis, the greater the opportunities. Whether your purpose involves developing new products, launching new services or exploring new strategies, dealing with change is inevitable. When embraced, change creates new possibilities. It opens the doors to growth and unanticipated gains. Often change is uncomfortable, but you can revel in the discomfort and use it to fuel your growth. Success or the feeling associated with "winning" is a result of stretching outside your comfort zone to the unknown, unproven, and uncharted. It requires an unwillingness to settle for the way things are and to constantly challenge yourself and those around you to take part, enable, and inspire change. The most important principle behind the Corporate Athlete is to build a community of global leaders that will continue to advance research, share stories, and amplify a growing dialogue on how individuals and businesses are redefining success to create sustainable value.

INSPIRING DISRUPTION

The success of any change comes down to your ability as a Corporate Athlete to inspire others with your vision of "disruption" and to get people motivated about the future, engaging them to become a part of the transformation. Take any industry, be it technology, education, biotech or environment, and fast-forward ten years—even if you are not there yet, the products underlying consumer behaviour and the market now rests on the 2030 point on the trendline, positive or negative. When change is framed as an opportunity to create a new possibility in a particular field, it leads to an exciting outcome. Within the realm of moderation, people become less resistant to supporting a disruptive idea, less fearful of uncertainty. The key to effective change is to clearly express the potential payoff or reward for each individual. How will they win? It also requires articulating a clear plan from point A to point B.

Most people do not change their behaviours but prefer to stay within their comfort zone and avoid unnecessary change. What holds them back? Old patterns and ways of thinking can keep you stuck in the status quo. A commitment to the way it used to be, how you used to do it, and how it has worked in the past restricts your capability to innovate and catalyze disruptive ideas. In order to grow, it is imperative to challenge every single old assumption. Rethinking the way you do things keeps you agile and relevant, and the same can be applied to an organization. As previously discussed, the challenge is to avoid complexity. Keep change simple and follow through with your idea to avoid fatigue.

If you fail, it is better to fail forward and learn from the mistake and move on. As the classic business phrase goes, it is not about how quickly you fail, it is all about the education, the takeaways, and the mistakes you hopefully will not make again. By "failing forward" when things are headed downhill, you learn and gain in the process. Brené Brown says failure is an imperfect word because it is never the end of the story if you are smart.[41] Failures turn into lessons, and lessons make you better and more likely to succeed the next time around.

CHALLENGE THE STATUS QUO

The question in any situation, whether it is work or athletics, is not are you good at what you do, but rather are you good enough to get better? In 2017, I had a chance to spend time with Jim Collins, the author of *Good to Great*. He gave a fascinating lecture at a biotech conference and famously noted a quote from his book: "Good is the enemy of great." According to Collins, "good" is the main reason things do not become great. Most companies do not become great because they are good, and it is easy to settle for good. Good breeds complacency. It creates a comfort zone that shuts down radical thinking and crushes creativity.

41 Brené Brown, *The Gifts of Imperfection: Let Go of Who You Think You're Supposed to Be and Embrace Who You Are* (Minnesota: Hazelden Publishing, 2010).

BREAKAWAY

DON'T OUTSOURCE YOUR HEALTH

"Everything around you that you call life was made up by people that were no smarter than you."
—Steve Jobs

In the midst of hitting their personal health and fitness goals, people frequently tend to lack personal accountability. Take a moment to revisit the principles of the world-class Corporate Athlete's High-Performance Trapezium discussed in Principles One and Four. Simply put, the foundation is based on holding accountability to oneself. Without putting in the work required, everything is destined to be structurally weak—it is called the High-Performance Trapezium after all, not an underperformance trapezium. Developing the methodology to build the personal metrics for personal success is a part of gauging your performance (extraordinary, above average, poor or nonexistent). Reaching your peak IPS as a Corporate Athlete is a nonnegotiable process, and it begins with you developing your own barometer for success and not aspiring toward someone else's.

Sometimes we fall into the trap of rapidly trying to achieve the outcome we desire, and it may tempt you to rely solely on others to help you achieve a goal—just look at the thriving, multibillion-dollar health and wellness industry. There are too many training plans, too many diets, and too many self-proclaimed experts. Sometimes it feels like the blind leading the blind when it comes to the health industry, and it is too easy to fall into this minefield of misinformed advice that is often too general to achieve great results and not specific enough to be tailored to individual needs. Even though working with experts can be a great thing, ultimately, you have to do what feels right for you, and paying into short-term gimmicks or riding the wave of quick fixes is never the answer. Health is an ever-changing concept you need to constantly monitor and calibrate. If you have the means to work with outside experts to supplement your accountability, make sure to have clear plans of action linked to the results you want to achieve. Ensure that you frequently review and find immediate resolution to issues as they arise.

As a Corporate Athlete, you get to drive the process, and it is an unwavering process when it is about you giving 100 percent to achieve your purpose. When working with others, you define the outcomes and milestones you would like to achieve. If something does not happen, it is up to you and your coach to confront it directly rather than avoiding reality or hoping for miracles. Inevitably, the work you put in helps to build resilience. You cannot rely on others to keep you in good health. Become your own expert, take control of your own fitness and health because it always comes down to you, the actions you take, and the goals you make for yourself.

Corporate Athletes must constantly challenge themselves and the organization's boundaries, encouraging employees to reinvent themselves. You do not have to be a genius to do any of this. I am a living example of how anyone can affect change in the world, even if the problem they want to take on is not their main area of expertise. When I entered into the biotech world, I had no technical training. I am not a doctor nor a scientist. But through time and perseverance, I have been able to learn about a particular area and continue to surround myself with people who help me along the way.

We are all capable of making more deliberate decisions based on our unique purpose and how we use it to address real, far-reaching issues. No matter what age you are, where you come from or what profession you are in, there are many different ways to contribute—it is just a matter of taking the first step.

Look at the examples of individuals—many who are young—who are making major differences today. Their genuine passion allows them to mobilize entire communities around the globe. They are also very much on the radars of other youth who are, of course, the future.

For example, many young people are already familiar with teen environmental activist Greta Thunberg. They may not have the same appreciation for climate change as Greta, but they are paying attention to her actions on social media. They know when she is protesting, and they are learning to understand why. They are watching someone not much older than themselves who is passionate about her purpose, and they are seeing the impact that can have on their own communities and beyond. Greta is one individual who, through her passion and her leadership, was able to organize an entire movement. She has an amazing sense of awareness. She is incredibly in tune with what is important to her. She also has the choice of whether she wants to act or not. She does not have to talk about climate change. She is a celebrity now, so she also could choose to monetize that and bask in all the glory, but she has chosen not to do any of that. She has chosen to be very deliberate about what she does and how she behaves. The combination of awareness and choice is driving the impact she is making on the world. Climate change is just one of the many different issues out there. Can you imagine everything else we can tackle just by learning from her example and thinking outside of ourselves?

Challenging the status quo can require you to break out of your comfort zone. In the previous principles of this book, I have given you the tools to start making changes inside your own home, in your workplace, and in your community. Now, I want you to think even bigger. Even before the pandemic, I had begun to identify certain industries disrupting the way we do things. But after this year, witnessing the changes in education, telemedicine, and even global innovation and cooperation that brought forward several vaccines against COVID-19 in a matter of months, just to name a few, we have had a front-row seat to our ability to adapt, pivot, and surmount challenges. Coming of age in a worldwide crisis has the potential to mature a generation, creating a renewed appreciation for community, cooperation, and resilience, a generation that believes in participation as opposed to idleness. The following five areas—education, life sciences, artificial intelligence, technology, and the environment—are places where you can start to find significant, long-term problems you can help solve by using your attention, your energy, and your purpose.

EDUCATION

The greatest challenges facing humankind—from international financial crises and climate change to the COVID-19 pandemic—tend to converge, intertwine, and often exceed our understanding. While many societies, groups, and people can adapt reasonably well to this new globalized environment, others have fallen behind and risk being overwhelmed by converging pressures. Nowhere does this discussion matter more than in the classroom. Unless we can educate our youth about the real-time effects of globalization or individuation and provide them with the appropriate tools and knowledge to succeed in the new world, the disparity between those who adapt and those who do not will continue to grow larger. This gap will hinder our progress toward a shared sense of human community and erode the ability of our new global society to remain stable and prosperous. We need to revolutionize the classroom by catering to the most basic of human instincts—a desire and willingness to learn—so we can ensure the education system of today can produce the problem-solvers of tomorrow.

How do we light a fire in the classroom and design a more creative educational environment? We start by reconstructing what it means to educate. Today's world demands a dynamic learning environment that challenges students with real-life obstacles and encourages them to collaborate and communicate with their peers in order to develop adaptive solutions. An innovation-driven economy begins by empowering our youth with knowledge and experiences that improve their understanding and encourage them to pursue personal growth. While there is no one-size-fits-all approach to this, there are a few steps that can offer a solid start.

For today's education to translate into enhanced innovations in science, technology, engineering, and math as well as an increased focus on entrepreneurial skills and social awareness, we need teachers and mentors to act as supportive catalysts and provide a self-driven learning environment for their students. In the often ideological-driven debates over education, it is easy to lose sight of what matters most: the everyday interactions that occur between students and teachers. Therefore, the first step is to create a dynamic curriculum that allows students and educators to share stories, experiences, and dialogue as part of the learning process. This provides students with a supportive ecosystem of teachers and mentors who are motivated to connect with them and guide their personal growth.

The next step is to allow students to apply what they have learned through studies, discussions, and experiences and develop an idea that is meaningful to them. Students then need to be equipped with the resources to see their ideas come to fruition.

Finally, we need to equip our children with the skills needed for tackling big challenges and developing innovative solutions. This is an area of focus for the nonprofit I cofounded, YELL Canada, the Young Entrepreneur Leadership Launchpad. YELL is a dynamic ecosystem that engages and connects parents, teachers, and the professional community at large to play a meaningful role in the education of tomorrow's changemakers. YELL provides school districts with a cutting-edge curriculum designed to align with the most innovative new programs being developed at the world's leading universities, so the transition from high school to post-secondary can be seamless. One aspect of YELL is the integration of the UN Sustainable Development Goals into the curriculum to explore issues affecting our youth as global citizens. By exposing them to these issues at a young age, we hope to get them thinking about the creative solutions that will leave our world a better place. The program focuses on the following concepts:

- **Agility and Adaptability**—Innovation by its nature is a process of trial and error, so we must encourage our youth to be unafraid of trying new things. One of the ways we have done this in the past is to have students write a failure résumé, where each student reflects on experiences where they felt they had failed and determine what they learned from each situation.

- **Critical Thinking and Problem-Solving**—We encourage teaching through the use of questions because when done effectively, students develop the ability to play with ideas, reason, weigh evidence, and communicate effectively. Students in YELL are encouraged to explore the lessons they are taught and to come to their own conclusions about an issue.

- **Collaboration Across Networks**—Through YELL, students from different schools across each district come together under one roof. These students come from varied backgrounds and in many cases, are from different cliques that would not normally interact with one another. Taking these students out of their comfort zone and having them work in groups is meant to expose them to different points of view.

- **Entrepreneurship as a Multidisciplinary Tool**—Students in our programs are encouraged to take an initiative in a multidisciplinary approach to explore issues and develop creative solutions. YELL becomes a platform for students to take all that they have learned in their high school careers and pull together knowledge from different subjects to successfully produce their innovative idea.

- **Bridging the Gap and Building Meaningful Relationships**—Throughout the program, the students get to interact with professionals, are encouraged to build meaningful relationships, and are introduced to the idea of professional networking. We are able to bring students and mentors together and build a strong community of like-minded individuals, looking out for each other, bringing out the best in each other, and pushing each other to set the achievement bar higher.

- **Self-Discovery and Leadership**—From Day 1, students engage in several self-discovery workshops to reflect on their personal experiences and become more mindful. Then, when students are divided into groups, they quickly realize leadership is not merely a title, but a position of power gained by setting a positive example for others.

- **Everyone Makes a Difference**—The program helps students recognize that every person has several distilling personal experiences that influence them in a meaningful way. Every person has a story that includes the important people who influenced them to become the best version of themselves. We encourage students to think about who those people are and how they helped shape them into who they are today.

As social beings, we require sincere interaction and collaboration to form our identity and facilitate personal and professional growth. At YELL, we emphasize this concept through mentorship programs that allow students to develop an idea and build it into a viable business venture while building lasting relationships along the way. Our approach to rethinking education and the classroom begins with a simple shift of power from teacher to student. Through active mentorship and collaboration, we create a self-driven learning environment that provides students with the skills necessary to succeed in the modern economy and the awareness that their personal growth and success is made greater by the success of others.

I encourage you to think about ways you foster a willingness and excitement to learn in the youth of your own community. How can you offer a unique contribution to this vital conversation?

LIFE SCIENCES

There has never been a more exciting time in the history of medicine than the period we are experiencing today. Everyone has a stake in this era. Thanks to the internet, marketing, patient engagement, and accessibility of information, patients are more educated about their ailments and the various diagnostics and therapeutic options open to them.

We are experiencing a time where the practice of medicine is highly integrated and efficient. Biotechnology is changing the prescription from a world struggling to meet the escalating health problems of an aging population to one that both treats the ill more personally and focuses on wellness by preventing or delaying the onset of disease.

The genomic revolution's greatest impact was on clinical development, where increased understanding of diseases enabled us to better selectively affect certain people, to exclude certain people from clinical trials, or to affect them in a certain way at a specific time. Such changes in clinical development allowed for greater efficiency. At OncoSec, our company focused on taking the fight to the tumour with an intratumoral immunotherapy. With our increasing understanding of immunology and the mechanisms tumours employ to subvert the immune system, combined with emerging data on the importance of "silent" tumour mutations and neo-antigens, we believe there is an opportunity to achieve the goals of "precision medicine" with treatments that are patient specific. Instead of developing personalized drugs, we focused on developing treatments that lead to a personalized response. We asked ourselves: What if the medical and scientific community shifts the focus away from the therapy itself as a means of achieving precision to achieving precision in the biological effects of the therapy?

The field of personalized medicine is still in the early innings of development with some notable successes already taking place but understanding the functional/clinical relevance of the human genome/DNA will continue to be key as this industry develops. Right now, we see three broad groups that play an integral role and will be impacted as this industry evolves: technology providers, who improve the sensitivity and the interpretation of data; pharmaceutical/diagnostics companies, who assimilate data to prevent and treat diseases; and end users, who represent the benefactors (patients and caregivers) of said technology.

Finally, the biotech/life sciences industry will not be isolated from major influences affecting the overall healthcare industry. The reality of de facto healthcare cost control may mean significant reduction in sales for individual drugs and that personalized medicine has an opportunity to make more efficiency in the market forces driving rising drug prices. Although the pharmaceutical industry will not be isolated from major influences affecting the overall healthcare industry, the reality is that personalized medicine has a large role to play in de facto health care cost control and it may mean a significant reduction in sales for individual drugs that do not take a patient and disease-specific approach. Science will continue to influence this equation but, meanwhile, payers may play a more activist role via reimbursement criteria or other means to keep costs manageable for patients.

What aspect of this ever-growing and evolving field is most interesting to you? What solutions can you offer with your unique skill set?

TECHNOLOGY

Technology is a broad industry with many moving parts but the aspect I am most enthusiastic about is how specific devices and applications can act as major disruptors. The development of the smartphone, apps, and chip/microprocessors that facilitate communications between hardware, software, and the underlying digital fabric is advancing progress across many industries.

At the time of this writing, the next wave of disruptions stems from the global rollout of 5G technology, the new generation of mobile broadband. 5G technology has the power to transform countries, industries, and societies all over the world, prompting unprecedented growth. 5G smartphones will allow consumers to download a movie in less than one minute, browse the web ten times faster, experience lifelike virtual and augmented reality, and stream 4K video the same way users stream audio today.

But while 5G smartphones will be remarkable, focusing only on the 5G smartphone user experience is limiting the technology's true potential. 5G's larger purpose is to be the underlying digital fabric connecting all elements of our modern world. In particular, look for 5G to fundamentally change the way our manufacturing, automotive, and healthcare sectors operate.

According to a November 2019 IHS Markit study called *The 5G Economy: How 5G will contribute to the global economy* commissioned by Qualcomm, a major chip manufacturer, 5G will contribute to $13.2 trillion in global economic output in 2035, supporting 22 million jobs that year. A Defense Department report from 2019 states, "5G will enable a host of new technologies that will change the standard of public and private sector operations, from autonomous vehicles to smart cities, virtual reality, and battle networks."[42]

The manufacturing sector also stands to be one of the biggest winners in this next digital revolution. Deploying private 5G networks will allow manufacturers to eliminate the cumbersome bundles of ethernet cables all over factory floors and ceilings that connect the machines wirelessly to the cloud. In fact, there is increased opportunity to make further data connection that allows for deep learning based on the data. As a result of collecting so much data from factory operations, companies will be able to use AI to increase productivity and efficiency. Everything from sensors and hand-held tools to assembly-line robots will be wirelessly connected, allowing manufacturing to reconfigure production lines more quickly and flexibly.

The automotive industry offers another window into the transformative potential of 5G for the entire value chain. The new mobile network will revolutionize how vehicles are built as automakers employ wirelessly connected production robots to work on car body construction. It will affect how cars are serviced as repair shops tap secure software updates for complex telematics systems. 5G also will influence how passengers enjoy the ride as its high speed and low latency allow for more seamless streaming of entertainment and other features that can be integrated into the consumer experience.

By deploying 5G vehicles, we will be able to communicate with one another through the exchange of complex data packets only possible through 5G. They will also interact with road infrastructure, traffic lights, drivers, and

---

42 *The 5G Economy*, 1st ed. iHS Markit, 2019, pp. 20-22, https://www.qualcomm.com/documents/ihs-5g-economic-impact-study-2019 (accessed on May 15, 2020).

even passengers through 5G-enabled smartphones. This next level of connectivity, combined with completely transformed navigation and mapping systems enabled by 5G, may help reduce collisions and save lives through enhanced sensors and vehicle safety, pushing down costs associated with maintenance, accidents, and insurance.

5G also will impact healthcare. Historically, challenges surrounding the tracking and maintenance of medical records have made the field of healthcare somewhat of a laggard to digitization. 5G's reliable, faster, and more uniform data rates will create a future of telemedicine where doctors monitor, treat, and predict health challenges remotely, delivering affordable, quality care right to the living room.

The pace of the 5G rollout is unprecedented in the history of cellular and the above are only a few examples of how it can be deployed. This technology opens up a great deal of opportunity as it is introduced throughout the world and provides individuals and entrepreneurs a catalyst for unparalleled growth by doubling down on the 5G future. How can you use it to best support your own purpose and/or advance the work of others?

ARTIFICIAL INTELLIGENCE

The term artificial intelligence, or AI, is often misunderstood. Many people confuse it with superpowered robots or hyperintelligent devices. Even top business leaders lack a detailed sense of AI. The term generally is thought to refer to machines that respond to stimulation consistent with traditional responses from humans, given the human capacity for contemplation, judgment, and intention. According to researchers Shebhendu and Vijay, these software systems "make decisions which normally require [a] human level of expertise" and help people anticipate problems or deal with issues as they come up.[43] Such systems have three qualities that constitute the essence of artificial intelligence: intentionality, intelligence, and adaptability.

In its present iteration, AI refers to technologies that can perform and/or augment tasks, help better inform decisions, and create interactions that have traditionally required human intelligence, such as planning, reasoning from partial or uncertain information, and learning. Over time, this definition will continue to evolve based on machine learning.

One of the reasons for the growing role of AI is the tremendous opportunities for economic development it presents. A 2017 PricewaterhouseCoopers project called *Sizing the Prize: What's the Real Value of AI for Your Business and How You Can Capitalise?* estimated that "artificial intelligence technologies could increase global GDP by $15.7 trillion, a full 14 percent, by 2030."[44]

There are numerous examples where AI already is making an impact on the world and augmenting human capabilities in significant ways. AI is not a futuristic vision, but rather something that is here today and being integrated with and deployed into a variety of sectors. This includes fields such as finance, national security, healthcare, criminal justice, transportation, and smart cities.

For example, autonomous vehicles equipped with LiDAR (light detection and ranging) and remote sensors gather information from the vehicle's surroundings. LiDAR measures how long it takes light to reflect back from

---

43 Shukla Shubhendu. and Jaiswal Vijay, "Applicability of Artificial Intelligence in Different Fields of Life," *International Journal of Scientific Engineering and Research (IJSER)*, September 2013, vol. 1, no. 1.

44 Anand S. Rao and Gerard Verweij, "PwC's Global Artificial Intelligence Study: Sizing the Prize," *PwC*, 2017, https://www.pwc.com/gx/en/issues/analytics/assets/pwc-ai-analysis- sizing-the-prize-report.pdf (accessed on May 15, 2020).

objects, so it can create a depth map of any space. Onboard computers combine this information with sensor data to determine whether there are any dangerous conditions, if the vehicle needs to shift lanes, or if it should slow or stop completely. All of that material has to be analyzed instantly to avoid crashes and keep the vehicle in the proper lane. Some of the latest generation of smartphones have incorporated LiDAR technology. Now, when you are shopping for furniture you can see exactly how it will fit in a room. Incidentally, NASA is also developing and incorporating LiDAR technology for Mars missions.

In the healthcare industry, AI tools are helping designers improve computational sophistication. While humans are capable of reading CT images and labelling healthy cells and growths that could be problematic, radiologists charge $100 per hour for this service and can only complete up to four patient files an hour. Applying AI to this common need could save time, energy, and a great deal of money.

With all the advances it is capable of making, AI also comes with an important aspect of accountability and a new sense of focus on purpose from ethical and moral implications. Who is responsible and who has the fiduciary or legal rights in an increasingly AI-driven world? If AI stimulates a reorganization of outdated bureaucracies, what leadership mindset should individuals adopt? Human choices about AI will determine the manner in which it is integrated into organizational routines. Exactly how these processes are executed and the leaders responsible for implementing these decisions need to be well informed to better understand the implications of AI, because they will have substantial impact on the general public for the foreseeable future. AI may well be a revolution in human affairs and become the single most influential human innovation in history. What challenges do you see in your own community that could be solved with AI?

ENVIRONMENT

Many researchers, engineers, and environmentalists are expressing deep concerns about changes in the overall climate of the planet. The root cause of the continuous rise in temperature of the planet is global warming. The concentration of greenhouse gases in the atmosphere has been artificially increased by humankind at an alarming rate over the past two centuries. The planet has experienced the largest increase in surface temperature over the last 100 years. Between 1906 and 2006, Earth's average surface temperature augmented between 0.6 to 0.9 degrees Celsius; however, the last fifty years saw the rate of temperature increase nearly doubling. Sea levels have shown a rise of about 0.17 metres during the 20th century. The extent of Arctic sea ice has steadily reduced by 2.7 percent per decade since 1978.[45]

We need to continue developing technology for alternatives to fossil fuels and switch to renewable-energy sources. Fossil fuels are being continuously used to produce electricity and other sources of energy. The burning of these fuels produces gases such as carbon dioxide, methane, and nitrous oxides, which lead to global warming. The usage of fossil fuels should be discontinued immediately.

Entrepreneurs interested in providing solutions to climate change must seriously pursue alternative energy sources, including wind, solar, biomass, geothermal, and hydro, as these do not produce any sort of pollution or toxic gases that can lead to global warming. They are environmentally

45 Terence C. Mills, "Modelling Current Trends in Northern Hemisphere Temperatures," *Royal Meteorological Society (RMetS)*, John Wiley & Sons, Ltd, January 26, 2006, rmets.onlinelibrary.wiley.com/doi/pdf/10.1002/joc.1286 (accessed on 28 December 2019).

friendly and pose no threat to ecological balance. Their high installation and setup costs may drive energy companies away from them at first, but in the long run, they are surely beneficial for everyone.

To counteract the medical hazards of global warming, it is essential to turn to a purpose mindset when considering energy sources. Everyone should be responsible about their decisions on energy conservation methods. Governments should devise and pass policies that encourage everyone, including energy companies, to use renewable energy instead of conventional energy. Nongovernmental organizations should distribute information to motivate people to use alternative sources of energy and discourage them from using fossil fuels.

Many developed countries are already generating huge amounts of power using renewables. These countries should extend their helping hand to developing countries to combat the negative effects of global warming collectively. Using renewable energy may be the most effective way to curtail the emission of gases, which play a major role in global warming.

What aspect of environmental changes interest you the most? How can your expertise be applied to creating solutions in this area?

COURAGEOUS LEADERSHIP

We have established that a world-class Corporate Athlete is not always operating in a peak training zone and courage may not necessarily be the first thing that comes to mind when running at a steady pace. Sometimes courage is the quiet voice telling you to run faster, or at the end of the day, "I will give it another go tomorrow." The principles of a Corporate Athlete provide so much context for leadership potential. Being in any position of leadership is an honour and privilege. With that, comes the responsibility to steadily challenge your boundaries, to stretch and grow and to constantly seek the next level of mastery as a Corporate Athlete. Being a Corporate Athlete conveys a duty to realize your fullest potential as well as the potential of others.

We live in a world that is constantly evolving. Most likely, you currently work in an environment that is advancing rapidly or perhaps the five themes presented in this principle have your wheels turning on the possibility of upcoming changes you can embrace or solve. You are either in growth or decay. There is no middle ground. As Theodore Roosevelt once said, "This country will not be a good place for any of us to live in unless we make it a good place for all of us to live in."

Courageous leadership necessitates you to take action. It invites you to pierce through fear, worry or doubt. A courageous leader stands in certainty and exudes self-confidence rather than arrogance, strength rather than control. These are times that require bold, confident, courageous leadership. Corporate Athletes who have the guts to take risks and break through barriers will be those who are most respected.

Demonstrating leadership courage can be uncomfortable. Whether it is having a tough conversation, making a contested decision or accepting criticism, it is based in humility and a never-ending desire to learn and grow. Corporate Athletes are courageous leaders who chart new frontiers and face adversity head-on. You put your most prized assets, your life, yourself, and your people, before profits. You are willing to be truly accountable to yourself, exemplifying authenticity, vulnerability, and transparency. You are continually growing and learning, always challenging the status quo, striving to reach the next level of mastery, and taking a vigilant stand for every person in your life to realize more of his or her potential.

As a courageous leader, you listen, invite candid feedback, and say what needs to be said. You give other people the credit and hold yourself to the highest level of accountability. You have the honesty to admit when you have made a mistake. You do not let your pride lead you in the wrong direction. You own up to errors and make things right. You are willing to embrace new ideas and challenge old assumptions.

RESILIENCE: HOW FAR WILL YOU GO?

"Float like a butterfly, sting like a bee—
his hands can't hit what his eyes can't see."
—MUHAMMED ALI

Like the butterfly, you begin your Corporate Athlete journey as a caterpillar, absorbing as much knowledge as possible, engaging in continual learning and improvement, reading, and seeking nourishment to help you grow. As you shed old paradigms and belief systems, you re-create yourself as you search for your truth. At some point, you shrug off the old in search of freedom from the past—old constructs, structures, and processes.

The shedding is not always an easy process as you battle with your ego, the establishment, or the way things have always been done. It can be painful. The emergence of the butterfly is resonant with the evolution and the realization of your potential, the promise of new life. Like the caterpillar, you must be patient as you surrender yourself to the process of transformation and the emergence of a more brilliant version of yourself.

How voracious is your appetite to confront your purpose with resilience? Are you willing to confront your resistance, feel uncomfortable, experience frustration, pain or failure?

Transformation begins with awareness of where you are currently and taking inventory of what you already have to achieve your purpose, evaluating your own high-performance trapezium and determining your ideal Joy Quotient. Once you have identified the gap between where you are and where you want to be, you ask, "What kind of Corporate Athlete will I need to be to get from here to past the finish line?" This is where you dive into the willingness and allow yourself to transform to be able to "sting like a bee" when needed. You stand in the field of becoming with radical honesty and you let go of what no longer serves you. In reaching forward and achieving the next level of success, you inevitably have to become someone different, someone more aligned with your purpose, running without effort in your ideal performance state, looking beyond the finish line.

Your purpose is the area in which you want to achieve greater success and, more importantly, more lasting fulfillment. To find your resilience, you need only look into your world and ask yourself, "What would I like to see happen? Where would I like to go? What would help me, others around me, my team, my colleagues get ahead and serve the greater good?"

When you challenge your strength and resilience, you reach deeper, see farther and unlock ingenuity. Sometimes tiny steps may not get you very far. Sometimes you must choose a leap sideways instead of forward. This begins with your willingness to make unreasonable requests of yourself. Corporate Athletes make it a point to keep stirring things up, alter routines, and break old habits. You push past current routines by zeroing in on the part that needs to be pushed further or recalibrating to help you achieve your purpose.

To continue evolving as a Corporate Athlete, you must say "no" to the drug of gradualness and get comfortable with taking calculated risks. Either we take a risk or we do not, and either change is measurable or it is not. The Corporate Athlete mindset is not about staying neutral, with the middle ground often placing you in complacency. Lasting change comes when you leap. Do you have what it takes?

□

WORLD-CLASS CODA

- Accept failures as a deep learning experience, as they are often the basis for the next big win. Pick yourself up, dust yourself off, and move on to your next project.

- Develop a plan to shake things up, disrupt traditional models, and spark more innovation and creativity. Be fearless in setting goals for yourself and your organization and be creative in your solutions to various challenges.

- Define what legacy you wish to leave. Identify ways in which you can deliberately and consciously build your legacy on a day-to-day basis.

CATAPULT FORWARD

- How well have you or your organization cultivated acceptance and appreciation of change? When did you last shake things up to stimulate a new level of creativity and innovation?

- How effectively do you exhibit the courage to be fearless? What fears, doubts or worries do you need to overcome to lead with more courage?

- What is your unique Corporate Athlete leadership footprint?

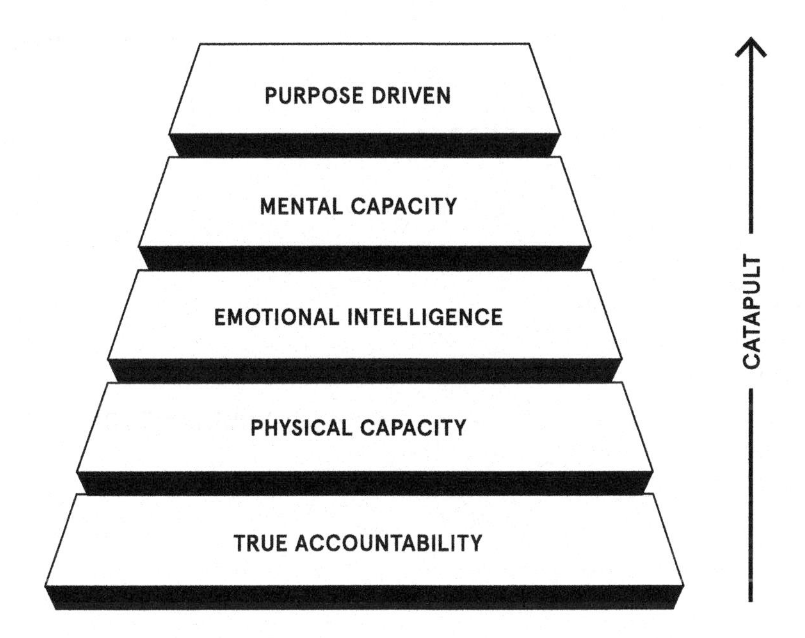

CAHPT and Catapult

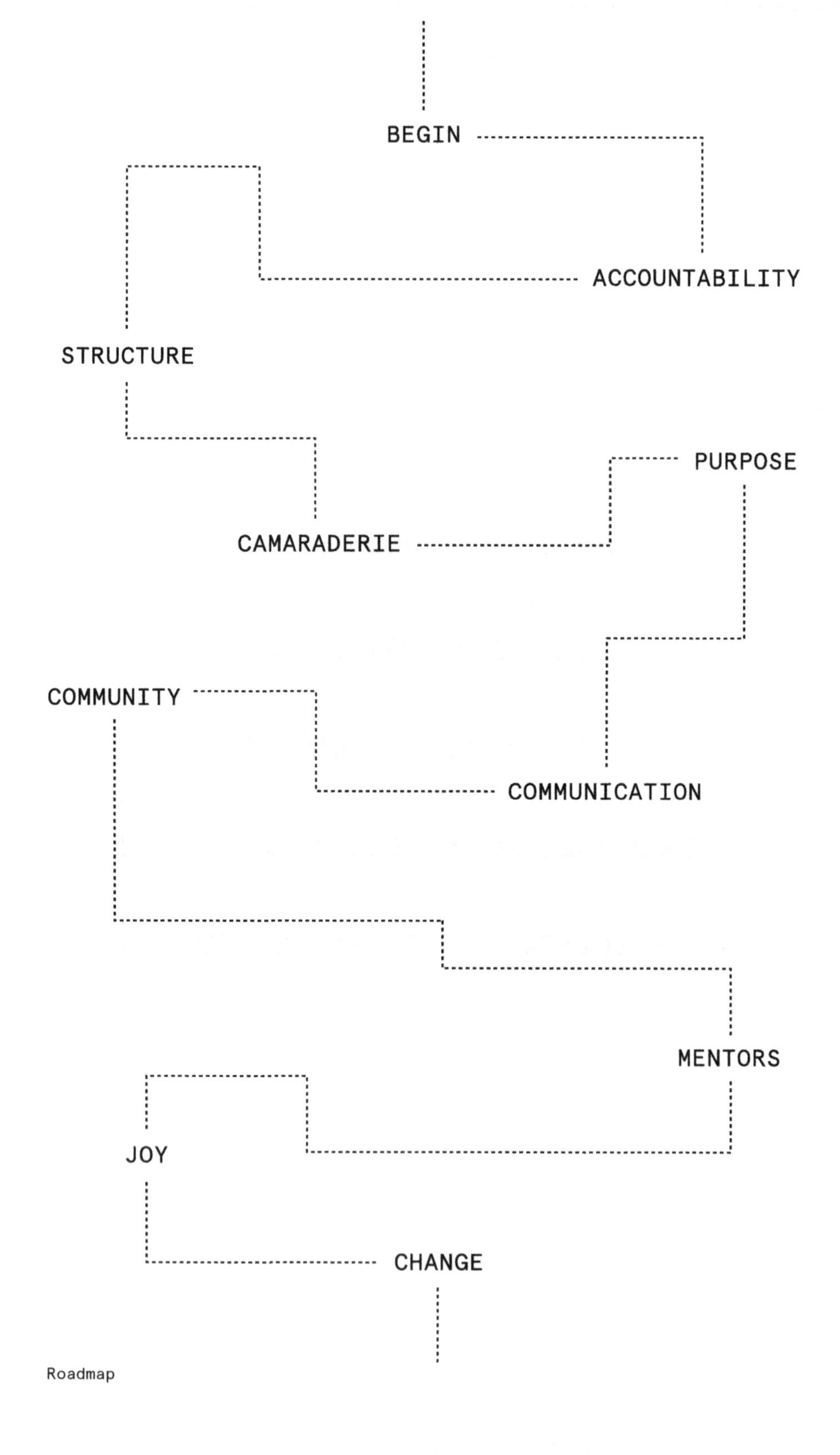
BEGIN
ACCOUNTABILITY
STRUCTURE
CAMARADERIE
PURPOSE
COMMUNITY
COMMUNICATION
MENTORS
JOY
CHANGE

Roadmap

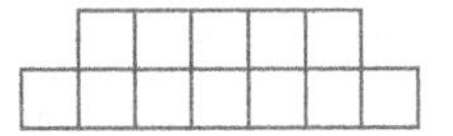

CONCLUSION

Every Saturday, I ride with a group called Breakaway Training. Rain or shine, through all four seasons, I am up before the first signs of light. I love the ritual of it—the steps that take me through the house as I get ready to go, a choreographed dance. A shower. My clothes. A glass of water. A snack. My keys. Go. As I slowly mount the bicycle in the driveway, my family soundly fast asleep in the home behind me, the leafy neighbourhood is still and quiet. I have a moment where the entire world is mine. As I start to pedal away, I ask myself a simple question: Why do I do this?

Why do I wake up every Saturday morning, get out of bed, and go biking with this group of people? I do it and all the other things I do every day because it is a component of my own purpose-driven life. I am trying to achieve 100 percent intrinsic motivation. For me, part of that has been about constantly surrounding myself with inspiring people who reinforce the principles I live by that keep me on the path I want to be on. I am lucky enough to be part of a dedicated and inspiring team of men and women who appreciate the significance of positivity as well as the camaraderie that comes with working in a peloton. Although many of us practice individual sports, when the squad is together, it is all about the team. It is never about winning or losing or success or failure. It is about supporting one another past the finish line. We know that requires helping one another, maintaining focus, prioritizing, having fun, and, ultimately, each of us understanding our own purpose and unique offerings as a member of the team.

In an ideal world, all entrepreneurs would have the same enthusiasm to take on any venture, completely certain it will bring us joy and fulfillment. Unfortunately, that is not always the case and inevitably there will be times when the work seems too daunting or even pointless. But when you truly know who you are, what you stand for, and why you do what you do every day—when you know your purpose—those moments will never derail your focus. As an entrepreneur, your purpose is your fuel. It defines and shapes everything you do. It makes you accountable to yourself first and, therefore, more prepared and ready to face the world.

By now, you know that requires establishing a strong foundation and structure, as it will give you the composure, stamina, and capability essential to maintaining your passion. You know it takes the action of opening the doors for others and constantly seeking opportunities to invest in one another's success. It takes recognizing when you are taking on too much or overcomplicating things. It means knowing how to effectively bolster your community and, perhaps more importantly, when to be still and listen.

You know that effective leadership as a Corporate Athlete means accepting the fact you will not always have all the answers and understanding when you need to reach out to experts or your mentors for help. It also means knowing how to spot opportunities to inject fun into your life in order to create the right energy of positivity in your environment. Even when you have reached the desired level of personal or professional success, it means remaining dedicated to growing, improving, and continuing to seek out solutions for the problems affecting your community and the world at large.

Above all else, being the best entrepreneur, leader, Corporate Athlete, and human being you can be requires understanding that your own authenticity and unique purpose are at the heart of everything you do. Becoming firm in your purpose frees you from uncertainty and ambiguity and helps to establish resilience. It empowers and motivates you. It reinforces your accountability to yourself as it fuels within you an overwhelming drive to act. Knowing what you stand for also encourages other people to get behind what you are doing, leading to a higher likelihood of success.

Having a purpose-driven mindset inevitably helps you distill your strengths and passions, which alleviates stress while simultaneously compelling you to do more with your skills, time, and energy. When you feel connected with the work you are doing, as well as the environment in which you are doing it and the people you are doing it with, your motivation is entirely intrinsic. You jump out of bed each day ready to take on whatever challenge might come your way because purpose generates joy.

Just as a child will not let outside stress or worry impact their ability to find happiness in all that they do, you too are able to funnel positivity into all that you do. Even as you take on more responsibility, you can continue to find fulfillment and new sources of motivation and inspiration as you work toward your desired outcome or objective.

While there are countless ways you can define it, purpose is not about words. It is about action and the discipline to commit and train your body and mind. When you know your purpose and trust in your own ability, you become accountable to yourself, which in turn allows you to elevate the people around you in the shared goal of stimulating innovation and creating positive change.

In many ways, the journey of making this book parallels the story within it and my personal mission toward my own purpose: to lead people and communities to pioneer indelible impact. I started this book with the intent of capturing my own principles without any idea of how to write a book. I had a few false starts and at the end, applied all of my knowledge and discipline to get it finished.

The day I pressed send on the final manuscript, I felt not only relief but also an overwhelming feeling that this was not about running past the finish line. The idea of a world-class Corporate Athlete mindset was not about winning or losing but about establishing resilience in any situation. This book has already delivered me someplace new. I am a different person and in a different place than when I first started. It has opened up all kinds of ideas and opportunities. The experience of making *Catapult* has changed me and is leading to my next adventure.

This is it. This is your life. The decisions you make every day, minute by minute, second by second, add up to determine how it will unfold. You get to choose what comes next.

Will you choose the Corporate Athlete path? Will you create your own set of principles? Will you tackle some of the challenges with the coming wave of innovation? Or will you follow the low-risk, well-trodden route travelled by everyone? The choice does not need to be radical. Today's youth and young entrepreneurs who will enter tomorrow's workforce will need to be more agile, displaying more stamina, endurance, and familiarity with the broader workings of the world to be able to find a niche that they can tackle. Will you invest the time and establish the purpose-to-impact plan in the pursuit of the changes you want to enact?

I often think back to Killarney Community Centre pool, to the coaches and fellow athletes who showed me the value of community and instilled in me the discipline of physical and mental training that reinforced my resilience. The Corporate Athlete mindset is a tool to guide you based on the growing economic diversity and increasing pace of change that we are witnessing in global business. This is a revolutionary time in our history, where the old paradigms about risk versus safety no longer exist. The Corporate Athlete needs to be able to bounce back quickly based on the foundation they have built for themselves. Sometimes, pursuing your purpose means going after challenges with the knowledge that the notion of stability is gone and the only thing you can be sure of is that the future will be volatile. For the Corporate Athlete, we are establishing the benchmark for leadership. The path you choose right now, at this moment, will determine how prepared you are to handle change when it comes knocking on your door.

Purpose reverberates beyond yourself and is continually impacted by outside factors. By thinking like a Corporate Athlete, it becomes about much more than simply doing a job. When you root your endeavours in your purpose, your work takes on new meaning and enriches and enhances all aspects of your life, providing you with the structure to be resilient against uncertainty. You become more thoughtful and proactive, which leads to deeper fulfillment and new levels of personal growth, enabling you to become a true agent of change. The world is waiting for you.

PUNIT
DHILLON

APPENDIX

PRINCIPLE TWO: NEED A COACH?

EMERGENETICS

Emergenetics is a unique assessment that provides an informative view of neural science and insights into seven key attributes of the way people prefer to think and behave. Drs. Geil Browning and Wendell Williams developed the Emergenetics tool to combine the core principles of effective learning, communication styles, and team interaction. On a practical level, the Emergenetics Profile represents a clear framework of easily recognizable and useful factors that apply to work, communication, and interpersonal relationships.

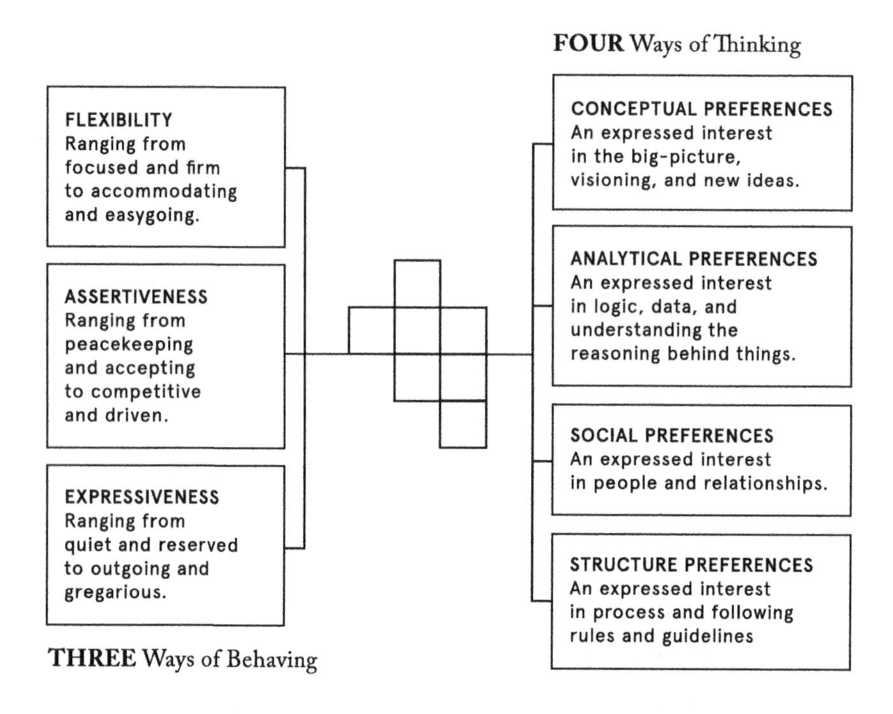

Behavior-Thought Confluence[46]

EQ-I 2.0 (EMOTIONAL QUOTIENT INVENTORY)

Emotional Intelligence (EI) is part nature and part nurture—it refers to a distinct combination of emotional and social skills and competencies that influence our overall capability to cope effectively with the demands and pressures of work and life. We cannot change our genetic makeup; however, we can become aware of what we have learned and we can choose to do better. Incorporating more than twenty years of research and development,

46 Adapted from Emergenetics International; R. Wendell Williams, "The Emergenetics Profile Technical Report." *Emergenetics*, Emergenetics International, Feb. 2018, emergenetics.com/wp-content/uploads/2020/10/Technical-Manual-February-2018.pdf.

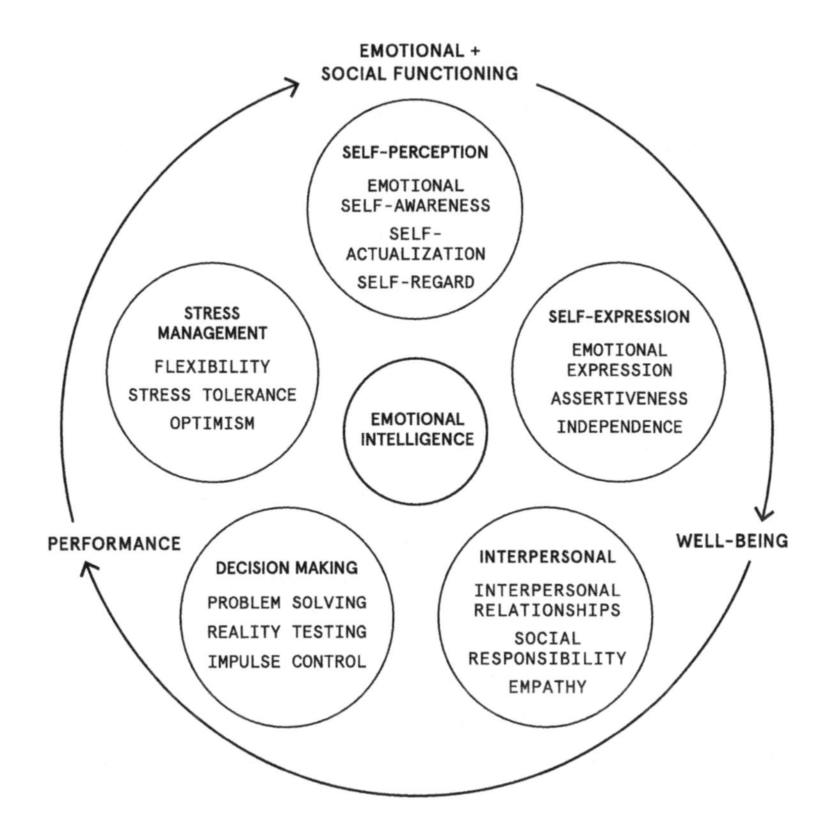

Emotional Intelligence Quotient[47]

the EQi is a psychometrically sound, validated assessment instrument that is applied to EI assessment and development at individual, team, and organizational levels. The EQ-i 2.0 model of emotional intelligence is based on fifteen competencies grouped into five composites: Self-Perception, Self-Expression, Interpersonal, Decision Making, and Stress Management. The idea behind EQi is to focus on these specific EQ competencies in order to

47 Adapted from the Emotional Intelligence Training Company Inc.; Ernest H. O'Boyle, et al., "The Relation between Emotional Intelligence and Job Performance: A Meta-Analysis," *Journal of Organizational Behavior,* 2010, vol. 32, no. 5, pp. 788–818., doi:10.1002/job.714.

make better decisions, communicate with greater clarity, increase our capacity to cope with stress, and build stronger, more connected relationships. The fifteen competencies, taken together, provide a total EQ. Everything is published in a final EQ-i 2.0 report, which also provides some initial interpretation of the fifteen competencies. A one-to-one debrief is always included with an EQ-i 2.0 assessment, and the results need to be interpreted by a certified administrator.

PURPOSE (MOTIVATION FACTOR)

From recent brain research, we know that our motivation is based on our individual needs and talents. In personal change and development processes, it is absolutely essential to focus on what drives you toward the goal and it is equally important to know what to avoid in order to stay motivated. With this knowledge, you can work determined and focused with your motivation and willingness to adapt. The purpose of Motivation Factor is to provide feedback on your current level of motivation, provide inspiration to increase your motivation capability and your intrinsic motivation, identify your top needs and talents, help you understand how to meet your needs and leverage your talents, and, finally, identify the level with which you associate the achievement of your personal purpose in a corporate setting within the organization's strategy.

The Pin Pointer Pro is based on the Hierarchy of Motivation. Recent brain research supports the theory that your ability to incorporate new learning and to manage change grows proportionally with how well you are able to take care of each level in the Hierarchy of Motivation. The Hierarchy of Motivation is based theoretically and empirically on positive and cognitive psychology, well-established motivation theories, and recent discoveries in neuropsychology.

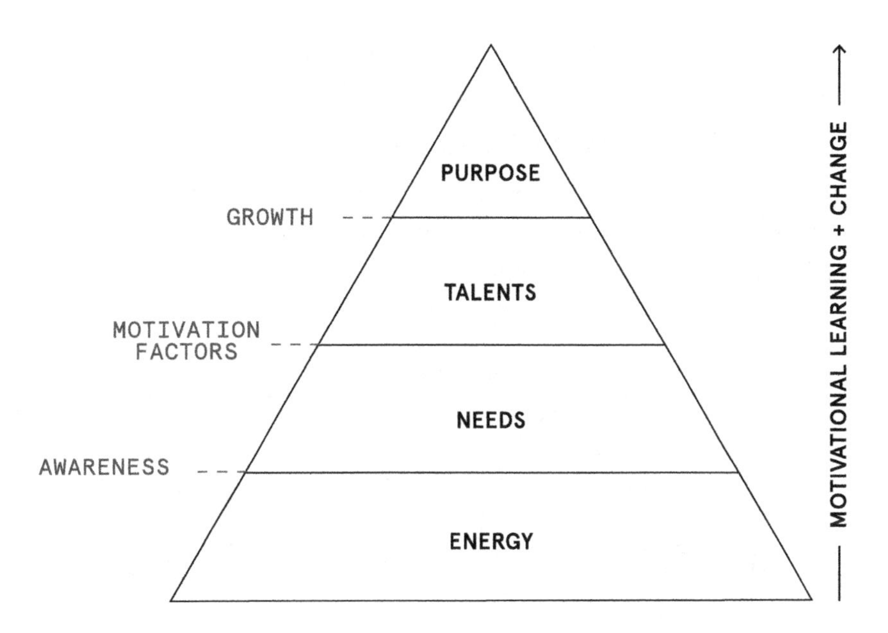

Hierarchy of Motivation[48]

ETHICAL LENS INVENTORY™

Our personal values guide the decisions we make every day, but sometimes it can be difficult to understand how we and others approach an ethical decision. The Ethical Lens Inventory™ (ELI) is a personal evaluation tool designed to help people understand the values that influence their choices. It identifies how they prioritize values when making ethical decisions. By understanding what values are most important to them and what values are most important to the other parties involved in an ethical situation, they can minimize unnecessary conflict, make better decisions, and live

48 Adapted from Factor Pin Pointer Pro; Jeffery Roy and Helle Bundgaard, *The Motivated Brain* (South Carolina: CreateSpace Independent Publishing Platform, 2014).

their values with confidence and integrity. This assessment was conducted on EthicsGame, a set of online tools designed to teach ethical awareness, critical thinking, and ethical decision-making. The underlying framework incorporates and compares deontological, teleological, justice, and virtue theories of ethics. The previous page has an illustration of the quadrants and values in tension that comprise the ELI.

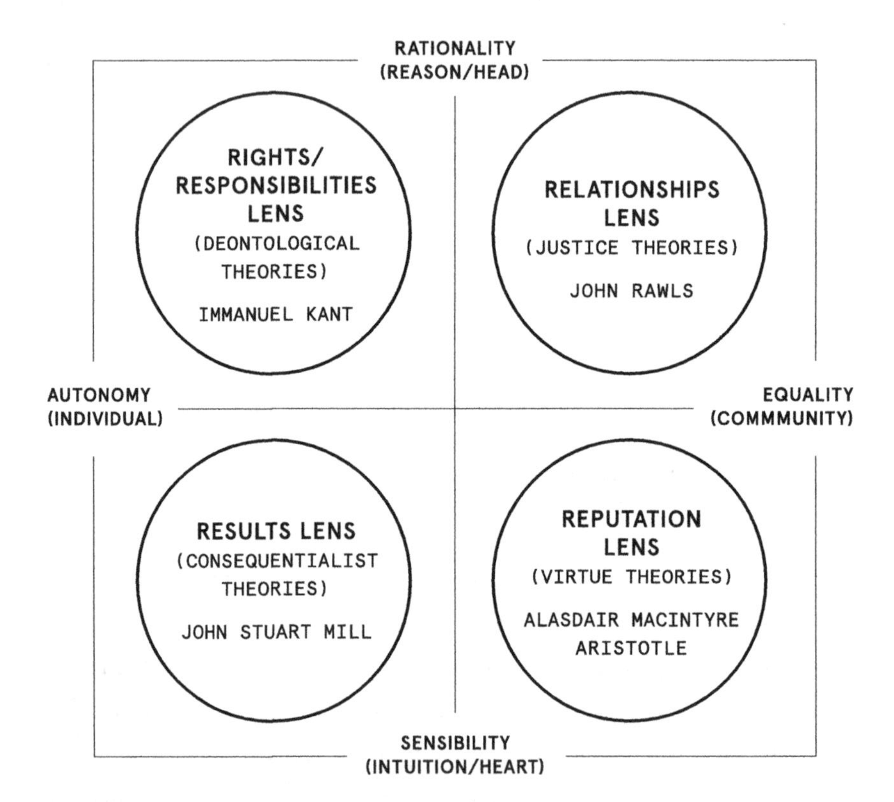

Ethical Lens Inventory[49]

49 Adapted from Ethics Game Press; Catharyn A. Baird and Jeannine Niacaris, *The Person in Community: Using the Ethical Lens Inventory to Enhance Your Personal Ethical Agility* (Denver: EthicsGame Press, 2010).

ACKNOWLEDGEMENTS

There is an African proverb that states, *If you want to go fast, go alone; if you want to go far, go together.* Nothing I have accomplished in my life would be possible without my community.

The book begins with a story of the final mile of an Ironman, but my personal story began while working at my grandparents' farm, picking blueberries every summer since the age of five. I soon made my way into the workforce as a gas station attendant, a concrete pump operator, a lifeguard, and a swim coach. I owe my Biji (grandmother) and late grandfather the utmost gratitude for the confidence they had in me and for providing the experience and work ethic that came from long days of working on the farm and managing all operations. I owe my parents, Jaswinder and Bhupinder, a debt of gratitude for making me work those tough jobs. Thank you to my two uncles, Avtar and Tawinder, for being amazing role models, forming the cornerstone in my upbringing. I have appreciated the labour of hard work from an early age and those jobs and people have made me who I am today.

Long after the hot summers spent pushing concrete wheelbarrows came to an end, my uncle, Avtar Dhillon, MD, made a bet with me that we could launch a new company together as a team. He envisioned the use of cutting-edge, novel molecules that could dramatically advance progress for several therapeutic applications with limited options, including the treatment of glaucoma. Avtar took me under his wing when I was just eighteen years old, nurturing and fostering my growth in the business world. The twenty-plus years I have spent working for him and with him and serving alongside him in various roles has given me the privilege of a seat at the table. This experience has undoubtedly informed the writing of *Catapult*. Thank you, Dr. Dhillon, for being an exceptional coach.

My family: my younger brother, Maheep, who is a constant reminder that there are no limits; my incredibly supportive mother, Jaswinder, whose strength is unparalleled; Nina, my beautiful wife, who is both an anchor and an undeniable force, obliterating any obstacles that stand in the way of our goals; my kids, Saiyrah and Reyha, who have provided new meaning to my life as they continue to teach me every single day and challenge me to be a better person. I am especially blessed and thankful to all of you who have been supportive of me as I have taken on the life of a Corporate Athlete. It is more than a little crazy at times and I will admit the balance has been unfairly skewed between work-self-life. Thank you to my extended family, whose hard work and experiences have given me the life I have today: my mother-in-law,

Gurmit Dhiman; my aunts who have been like second moms to me, Diljit, Bim, and Roni; and the brothers and sisters and family I have gained through marriage. Thanks for always being there to support and celebrate.

I am profoundly grateful to Olivia Yung and her wisdom and diligence, taking the beginning of an idea, working with me, and turning it into this book. My first draft was a 120,000-word mess, the product of a first-time author. Her sustained attention, coaching, and copious editing produced the book you have in your hands now. My gratitude to you friend —thank you for agreeing to help me with this outlandish project after I reminded you how fun it was when we worked together on that play you wrote over twenty years ago while we were still in high school. Thank you for applying your above-average IQ and writing skills to test every assumption and scrutinize every sentence, never allowing this book to lose its soul. Your uncanny ability to make things happen shows me that true collaboration is priceless. You have been an indispensable partner.

I would like to express my special thanks to the entire team at Scribe Media: Emily Anderson, Jericho Westendorf, Kacy Wren, Rachael Brandenburg, John van der Woude, and the members of your team who helped guide me through the process of writing and publishing this book. Rachel LaBar, you were right—at some point, we just have to put the pen down.

I deeply appreciate the work of my PR team, the Very Polite Agency, especially Andrea Mestrovic and JJ Wilson for knowing that I had this book in me, sharing my vision and helping create a book that educates and challenges the next generation of young entrepreneurs. Thank you for deploying your entire team for this project. The design team, led by Dylan Rekert, has created the most visually stimulating book and brand. With their guidance, I have been given a web platform that extends beyond this book.

The research and analysis provided by Justin Noppé and Reena Sandhu, PhD, has been invaluable to the final product—thank you both for lending your brain. This team has made this dream of publishing my ideas, management experience, and training philosophy become a reality.

I relied on a group of essential friends and advisors who offered wisdom, judgement, and sympathy throughout the writing process. Among the many, I must single out:

Judy Brooks, my touchstone for the last decade, who offered me invaluable perspective on purpose and immediately understood the why of this book better than anybody.

Adil Daud, MD, the scientific founder of OncoSec, whose unwavering belief in me and willingness to champion my cause has given me the confidence to think outside the box in order to get results.

Jim Heppell, my first mentor and dear friend who showed me the ropes and stuck by me as I grew from a novice student-athlete to accomplished Corporate Athlete.

Annalisa Jenkins, MD, a dynamic businesswoman with a heart of gold. Thank you for providing me with much-needed tough love and for holding me to the highest standards.

Sharad Khare, a consummate yes-head to any idea that I have ever proposed.

Joseph Kim, PhD, my former supervisor and CEO of Inovio Pharmaceuticals, who has the most admirable work ethic and drive, surrounding himself with and learning from eminent individuals from different fields.

Anil Kukreja, who has one of the sharpest managerial minds out of anybody I have ever known. Everybody should have a wise friend like Anil.

Iacob Mathiesen, PhD, a close friend, whose deep love of nature and simplicity carries over to his work in his field and who, from Day 1, has challenged the assumptions of this book.

Kenny Santucci, for always offering encouragement and insight in your no-bullshit way when I needed it the most.

Jason Sarai, thank you for protecting me from myself.

Patricia Tu, for keeping me accountable to my ambitious nature and immediately recognizing the scaffolding that CAHPT brings.

The catalyst for this book, putting these thoughts on paper, was born out of a discussion with Jason Sarai and Curtis Christopherson in Toronto in October 2019 after we thought we ran into Wayne Gretzky in the lobby of the Ritz Carlton. Thank you both deeply for your help in this process. I drew on the skills, life experiences, and relationships I have with many great people to turn these ideas into a book.

My close friends: Tu Diep, my best friend since age thirteen, whose loyalty and steadfastness give me the confidence to take big risks; Winnie Lam, an amazing athlete, my biggest champion and sister, who pushes me to my limits and is never afraid to be unapologetically honest.

The YELL Canada organization, students, and community, including Sarah Lubik, PhD, the first director of entrepreneurship at SFU, fellow board member, and overall uplifting innovator. It is an honour to learn from you and be a witness to your invaluable insight. You are an inspiring dream facilitator for the next generation. I am grateful to call you a true friend. Amit Sandhu and Rattan Bagga, my brothers first and foremost, I am constantly in awe of your values and your dedication to making big changes in the community and beyond. I am proud to call you my colleagues and, most of all, my closest friends. With your unwavering support and encouragement, there is nothing I cannot achieve.

My coaches, who elevate me in many different areas of my life: Felipe Loureiro, my triathlon coach for the last decade, who has endured all of my whiny complaints and celebrated all of my incredible victories. Paul Chung and Minh Nguyen, my athletic trainers who have been able to push me to be physically and mentally unrelenting. My swim coaches and other inspirational trainers: Andrew Currie, Rob Moretto, Paul Moretto, and Paul Zelinski, as well as all of my other coaches from whom I have learned a great deal over the years.

My incredible mentors, colleagues, supporters, and believers in my abilities throughout my career: Jawahar Lal Bagga, Riaz Bandali, Reni Benjamin, Jim Demesa, MD, Bob Goodenow, PhD, Paul Grayson, Bob Hawk, Bernie Hertel, Peter Kies, Val Litwin, Riaz Meghji, Gaetano Morello, ND, Kewel and Dave Munger, Pamela Munster, MD, Michael Murray, ND, Terry Norchi, MD, Daniel O'Connor, John Rodriquez, Noam Rubenstein, Paramjit Sandhu, Vijay Samant, Chris Twitty, PhD, and Michael Vasinkevich.

Special thanks to the mentors at Brew One, including Sharon Duguid, Mary Prefontaine, and the fellow Brew brotherhood, who created an environment of trust that encourages us to challenge each other to live in purpose. You live and breathe the principles of this book and I am proud to be in your company.

The team at Torpedo Publishing, who worked tirelessly to breathe life into the manuscript, turning it from a concept to a real, tangible entity. This team has made this dream of publishing a book become a reality. I am truly indebted to your commitment, your vision, and your guidance. It has been the best adventure.

I also want to take a moment to recognize the following people, who have been supportive in every interaction we have ever had and, in many cases, opened new doors for opportunities and personal growth. You are incredible role models in this life. I could not have asked for more amazing cheerleaders to enable my purpose: Marco Amselem, Mona Ashiya, PhD, Sandra Aung, PhD, Jon Azen, Pavel Bains, Chris and Kristen Berg, Kyle C. Bisceglie, Orville Bovenschen, Devon Brooks, Maggie Campbell, David Canton, PhD, Justin Cha, Amy Chan, Coley Chavez, Shannon Chen, Chong Chin Cheong, Linda Chow, Wilson Chow, Linda DeLay, Ben Dhiman, Zaheer Dhruv, Leslie Ellis, PhD, Phil Feng, Sharron Gargosky, PhD, Gurdeep and Kathy Gill, Troy and Jane Gindt, Shawn Goyal, Tom Grant, Ian Hesselden, George Hincapie, Cheryl Hsu, Jaeger Hunter, Vina Gandhi, Deborah Geraghty, PhD, Rob Gilbert, Karlene Karst, ND, Thomas Kim, Vik Khanna, Ng Tee Khiang, Neeti Kukreja, Jessica Lee, Gabrielle Malette, Mary Marolla, Dan and Jessica Mah, Paul Melo, Valerie Morrison, Amin Nathoo, Tahmeena Nizam, Catherine Ngo, Cynthia Ortiz, William Pak, Shane Poppen, Derek Randall, MD, Kym Randall, Sean Rathbone, Lyndon Rosario, Joyce Rosario, Mr. and Mrs. Rosario, Daniel J. Ryan, Niranjan Sardesai, PhD, Lenny and Sally Sciarrino, Jennifer Tang, Greg and Denay Trinidad, Praveen Varshney, Luke Walton, Jon D. Williams, and Tracy Wong.

I am humbled to be in your company.

ABOUT THE AUTHOR

Punit Dhillon is an unrelenting entrepreneur focused on community, challenging the status quo, and a commitment to excellence.

Specializing in biotech and health sciences, Punit is the cofounder of OncoSec Medical Incorporated (NASDAQ: ONCS), a biotechnology company pioneering new technologies to fight cancer, and served as president and CEO. Prior to cofounding OncoSec, he was vice president of finance and operations at Inovio Pharmaceuticals, Inc. (NASDAQ: INO).

Currently, he is the chair and CEO of Skye Bioscience, Inc., a public pharmaceutical company. He also serves on the board of directors of other life science companies and acts as an advisor and consultant in several capacities to different life science organizations.

Punit is the creator of Torpedo Publishing, a publishing imprint serving to connect, inspire, and educate future game-changers looking to positively impact the world. Torpedo embraces a philosophy of change and global impact backed by science, sport, and innovation.

In his spare time, he is an endurance athlete and active runner, cyclist, and swimmer. He lives in San Diego with his wife, Nina, and their two daughters.